CROSS-COUNTRY COURSE
DESIGN AND
CONSTRUCTION

CROSS-COUNTRY COURSE
DESIGN AND
CONSTRUCTION

The Essential Guide for Course Designers, Builders and Competitors

Mike Etherington-Smith

J. A. ALLEN · LONDON

ISBN 0 85131 844 4

J.A. Allen
Clerkenwell House
Clerkenwell Green
London ECIR OHT

J.A. Allen is an imprint of Robert Hale Limited

The right of Mike Etherington-Smith to be identified as author
of this work has been asserted by him in accordance
with the Copyright, Designs and Patents Act 1988

British Library Cataloguing in Publication Data
A catalogue record for this book is available from the British Library

Photographs by the author except for the following:
half title, frontispiece, pages 1, 15, 35, 79, 129 and 168 by Kit Houghton,
page 69 (bottom) courtesy of John Britter Photography

The photo on page 167 is courtesy of Barriers international Ltd.,
designers and manufacturers of Frangible Pin fences.

Illustrations of the CCI*** on pages 151–162 are reproduced with thanks to Blenheim Horse Trials.

All line illustrations by Christine Bousfield

Edited by John Beaton
Design and typesetting by Paul Saunders
Map on page 39 created by Rodney Paull

Colour separation by Tenon & Polert Colour Scanning Limited, Hong Kong
Printed by Kyodo Printing Co (S'pore) Pte Ltd, Singapore

CONTENTS

NOTE ON MEASUREMENTS
Throughout the book metric measurements are used and the
approximate imperial equivelant is given.

·

PART ONE

Introduction

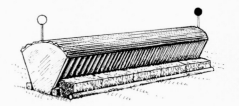

'It is our job to produce horses and riders and give them the opportunity to progress up the ladder'

THE DESIGNER'S RESPONSIBILITIES

THE AIM OF THIS BOOK is to give useful guidance and hints to all involved in the design and construction of cross-country courses, irrespective of the level and regardless of whether one is a professional or simply an enthusiast.

It is fair to say that over the last twenty-five years there has been a marked change in the courses that we see nowadays. The fences are generally more imposing, the materials are bigger, and the courses are undoubtedly more technical. This has come about for a variety of reasons – competitors and horses are better prepared and educated, particularly at the top end, there are more competitions, natural evolution of a sport, and, of course, the natural desire to improve.

The responsibility of course designers has never been greater and their profile has never been higher. Competitors quite rightly expect a good product when they go to a show, whatever the level and whether it be an event or a hunter trials. They need to know that a course will be built to the safest possible standard with good attention to detail meaning that we must all constantly ask ourselves whether we have done our best, and if not, why not. We are all accountable for what we produce, must accept this, and be fully aware of our responsibilities. We also have to try very hard to be right before the competition; any fool can tell you afterwards whether you have done a good job!

At the top end of the sport there is also the question of whether designers are providing enough of a test for the top riders at the highest level three-day events – the standard is so high that virtually whatever is produced on a course the top riders will answer the questions and be inside the time. Are we satisfying their needs? There are no easy answers but we must all constantly take stock of what we are doing, why, and whether the results are what we are looking for.

Hopefully there are some useful guidelines here for all involved in our sport. Safety has always been, and must always remain, the number one priority but there are always new things to learn and experiences to share for the common good and we need to all work together to improve our skills.

Course designers, and indeed builders, will all have their own style and this is to be encouraged. Courses would be very dull if they all looked the same and contained the same types of fence but it is often possible to identify the designer when walking a course: the type of questions, the flow, the distances, and the look of the fences themselves. The trick is to study all courses and take the good points from them, discarding the rest. All designers have good and bad points and we all make mistakes. However, we must learn from these mistakes and ensure that they do not happen again – there is always room for improvement!

THEORY OF COURSE DESIGN

What is a cross-country course? What are we trying to achieve and how are we going to achieve it? Simple questions with not so simple answers.

A good cross-country course is created from a cocktail of ingredients. It is about producing a series of jumps appropriate to the level of the competition that piece together and relate to each other to form a picture or a jigsaw that will allow the good horses and riders to come to the top without giving the not so good a bad experience. All the pieces of the jigsaw need to fit together in such a way that the end result is an entity in its own right; a course is not simply a series of fences put together in a random way; there must be a purpose, a goal, an understanding of what one is trying to achieve and how to set about achieving this goal.

As course designers we are looking to create courses that horses and riders will enjoy and benefit from. It is our job to produce horses and riders and give them the opportunity to progress up the ladder and part of this responsibility is to produce courses of the standard appropriate to the level of competition and not to push the boundaries beyond that particular level.

At the lower levels the emphasis will be more on education of horse and rider so that they have the chance to learn about each other whilst developing confidence and learning about the requirements of cross-country riding. The higher up the ladder they progress the more the balance turns to asking questions rather than being purely educational but we must always remember that we should be giving the horses and riders the opportunity to show what they can do and are capable of rather than the opposite. This is a very fine line but the difference is simple – one is positive thinking, the other is negative.

At the risk of stating the obvious any course must have a beginning, a middle, and an end. We need to think how we are going to start the competitors off so that they have every chance of getting into the course with confidence and in a good rhythm; we need to work out where and how we should ask the questions, and we must then finish the competitors on a good note so that they are ready for, and looking forward to, their next competition.

A course should have a good 'flow' and 'balance' to it. It should have a good 'feel' with a variety of fences and it should use the land and the natural features as well as possible. Imagine riding a horse in a good canter around the proposed route of the course without any fences on it and then ask yourself 'how will it feel?' Does the ground come comfortably to the horse or does it twist and turn too much? Will the riders have the opportunity to get their horses into a rhythm and then stay in it? If the answer is no to these questions then it needs a revisit. This is what is meant by the 'flow'. The 'balance' is the relationship between the fences themselves – where the questions are asked, how they are asked, and the consistency of the level of difficulty.

Every fence must have a reason for being there. It must serve a purpose as part of the overall picture and contribute to the course. If there is no reason or justification for a fence it should not be there. Also, and very importantly, let the courses design themselves. Do not feel obliged to have a particular type of fence for the sake of having one on the course, for instance, if there is no suitable place for a coffin fence, do not try to fit one in – it will be inappropriate, out of context, and will always be unsatisfactory.

To produce good courses a designer must have an understanding of how all types of fences and courses will ride. He/she must have a feel for how horses actually work physically, how they cope with differing types of terrain, how they jump the different types of fences, and how they cover the ground. A course designer must spend a huge amount of time just watching horses, the good ones and the not so good, on course and over fences, observing how they react and cope with differing questions and different terrain, in order to develop this understanding since, without it, it will be extremely difficult to produce a good cross-country course of any level.

At the lower levels we are looking to educate horses and riders, to see if riders can control and judge their pace, are able to set their horses up to jump simple questions and present their horses correctly to the fences. Simple questions on a turning line that allow the horses to be ridden in a good and sympathetic way without being pulled around, holding a line through two or three gently angled elements, jumping straightforward coffin type fences, combined with plenty of 'run and jump' fences, all these will develop the rider's skills and the horse's knowledge and confidence. We are wanting the

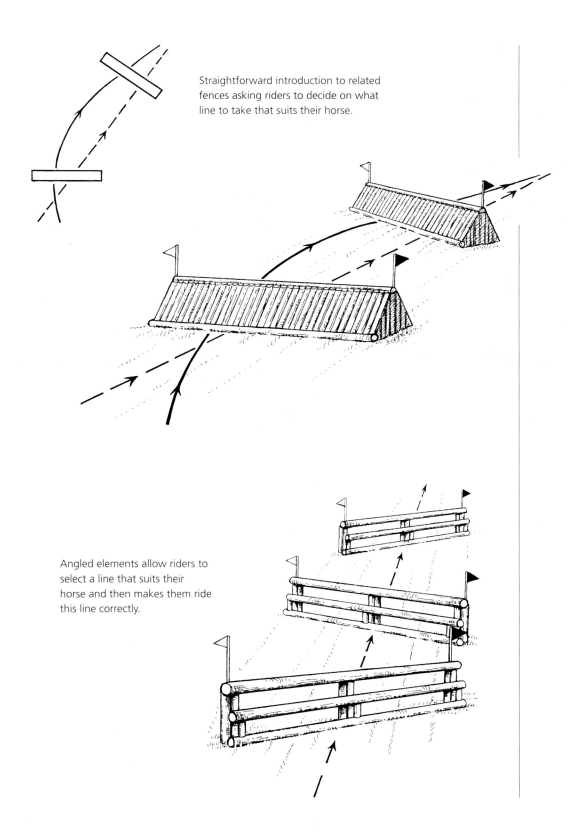

Straightforward introduction to related fences asking riders to decide on what line to take that suits their horse.

Angled elements allow riders to select a line that suits their horse and then makes them ride this line correctly.

Simple coffin fence with inviting profile on each element which encourages the horses to jump it well.

rider to learn about distances in related fences and how and where to present his horse to the fence such that it is easy for the horse to cope. Riders must learn and then recognize at this level what are good distances for their horses and what line to take through combinations; they must learn when and how to set up for a fence so that their horses are given every opportunity to negotiate the fence correctly; they must be given the variety of questions that will teach them and form the foundations for the type of questions that they will meet as they progress upwards.

At the higher levels the questions are inevitably more difficult, more technical, and more sophisticated. We are asking more of horse and rider; fences are bigger, angles increase, fences become narrower, expectations are greater. The emphasis moves towards the asking of questions rather than just education; there are less 'let up' fences, and the 'intensity of effort' becomes more noticeable.

It is a good plan to try to keep the fences coming regularly. We want to allow the horses to get settled into a rhythm and demonstrate their skill and ability and any/all questions should be well thought out and spread throughout the course. The frequency of the fences helps to keep the horses interested in their work and, more especially at three-day events, helps to stop them feeling tired. Long periods on a course with no jumping can allow a horse's attention to wander when ideally we want to keep them concentrating.

We must also be conscious that there are inevitably hold-ups on courses from time to time for a variety of reasons – fences broken or damaged, or perhaps a fall of some sort. This means that when planning a course there must be what we call 'stop' fences which are straightforward fences before which horses can be held while the problem further along the course is being sorted out; these fences need to be straightforward so that the horses have the opportunity to get going again over something relatively simple and there would normally be three or four such fences spread around a course. The first 'stop' fence should normally be at about fence 3 or 4 – sod's law says that there needs to be a hold just after a horse starts!

We must ask ourselves where are the questions? what are they? have we prepared the horses and riders for them? are there too many? are there enough? are they fair? are they in the right place? how many drops are there? and so on. The questions asked should be consistent in their degree of difficulty and no one fence should be designed to be more difficult than any other or exert more influence on the competition. Also, avoid having big drops at the end of a course when the horse's legs will be tiring.

Having the opportunity to ride some very good horses at all levels is undoubtedly a huge benefit. Equally, riding a number of not so good ones is also immensely beneficial in gaining an understanding of how horses cope with different types of terrain – if the less gifted ones can cope, the rest should find it easy.

It is also important to design out grey areas if possible. Recognize that others will be judging the fence and they will not know what is going through the designer's mind. Designers can make life easy or difficult for fence judges and Ground Juries. Often there is a fence on a course that can cause confusion to the judging of it because competitors are given opportunities for 'gamesmanship'. An example is at a combination fence where a decision has to be made as to whether a horse incurred a run out or not; was it avoiding the obstacle to be jumped or was the rider deliberately doing this in order to avoid getting penalties? Design if at all possible in such a way that there can be no doubt.

All designers must thoroughly understand how fences are built and what materials are good (and those that are not good). A designer must work closely with the builder, respect his craft and skills and listen to his advice. A course evolves from a successful partnership between designer and builder to reflect the character of both; the building team must feel involved, taking pride in their work rather than just doing another job. As with most things in life, it all boils down to team work, effort, and talent.

All designers have their own style and signature. One useful tip, more applicable at a three-day event, is to try get the riders behind the clock in the first third of the course so that they then have to think where and how to

By having the unjumpable rails between the two parts of the second element of a combination it makes the judging easier should a horse run out at 'B' on the quick route because the rider will have, in all probability, to cross his tracks to get back to the alternative.

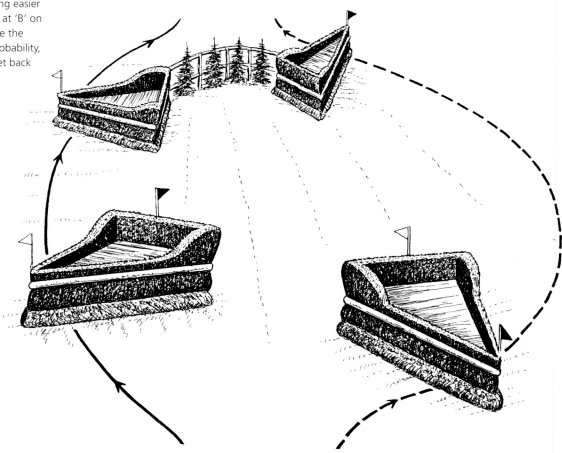

make this time up. This is the classic situation where 'theory is great, reality is very different' because it does not work very often. However, the principle is a good one in much the same way that it is good to make riders have to make regular decisions based on how their horse is going – get them to think their way around a course. With the shortening of the minimum permitted distances at three-day events but no reduction in the permitted number of jumping efforts, time will become more of a factor. This is a good move forward. It will help designers relate fences to each other better, it will hopefully make the time allowed more difficult to achieve, and it will reduce some of the long stretches on courses without any jumping.

In planning any course the required distance is obviously part of the equation. How to achieve this distance in the optimum way is what we all constantly strive for. Time is part of the test but when wheeling a course it is essential to be fair and take an achievable line that a good, well ridden horse will take. Do not cheat! Remember, the cross-country phase is part of an all round test, it is about achieving a standard, not about pushing that standard beyond where it should be by wheeling the course too tightly, thereby making the speed faster than laid down. Also, remember that the spread of a fence is distance covered and needs to be wheeled.

FENCES AND COURSES IN GENERAL

As the route starts to take shape we can begin to look at the fences in principle. In reality, if the route is good most courses design themselves. The temptation is always to try to get too clever or elaborate and as a consequence it is too easy to lose sight of what we are trying to achieve by over-complicating the issue. The KISS (Keep It Simple, Stupid) principle is undoubtedly to be recommended!

It is important to remember that, whatever the competition, a course is about achieving a standard. If, for example, the design is for a novice course, then make sure that it is a novice course, not too easy (i.e. virtually a pre-novice), nor too difficult (verging on intermediate). Competitors enter a competition at a level and expect to compete at that level, no more, no less, and were they all to jump the course clear (which they will not) then it shows that they are all up to the standard of that particular level; it does not mean that the course is too easy if it is right to start with.

In eventing all courses are geared to the highest international level i.e. four star. The levels below that need to form a ladder of progression so that horses and riders have the chance to work their way up the grades. This ladder is crucial if we are to produce horses and riders for the future. One star must feed two star which must feed three star which in turn feeds four star level competitions and this progression also applies to one-day events. If the steps on the ladder get too big because a standard is compromised or softened there is a greater chance of horses and riders ending up over-faced and lacking confidence; the consequences are obvious.

It is worth dismissing a misguided belief at this stage that small equates to easy. This is simply not the case. Fences that are small are not necessarily easier than bigger ones. We can all produce a 1.10m (3ft 7in) course that is unjumpable but this is not a clever move, nor does it prove anything. Any horse at the level at which it is competing should be able to jump a straightforward fence of maximum dimensions; if it cannot then it is not ready for that particular level.The degree of difficulty should come with the technicality of the course, the relationship between the fences, and their siting, and this is where the skill of the designer comes in, pitching the level of difficulty just right.

These photographs illustrate that easy need not necessarily mean that a fence is small but any horse at any level should be able to jump a straightforward fence of maximum dimensions

At the lower end of the sport the emphasis is on education and introducing horses and riders to the type of questions, albeit in a straightforward way, that they will meet as they progress through the grades. Get them jumping angled or curved rails and simple coffins, ditches, and corners. Increase the horses' and riders' expertize in such a way that they are not horrified when they meet a fence at a higher level. At the lower levels, when a question is asked then consider giving a straightforward fence to follow as a confidence booster; if there is a ditch that perhaps looks a little imposing then give the horses an inviting one earlier in the course as a preparation – remember we are here to produce horses, not write them off. At the higher levels there will inevitably be more technical questions and less 'let up' fences. However, the same principles apply.

We want to try to produce courses that do not destroy a rider's confidence when he or she first walks round. A rider should start off with a feeling of 'if all goes well I could perhaps be in the money'. As designers we want them in a positive frame of mind which will also transmit to their horse; we want them eager and enthusiastic as well as slightly nervous. Ideally we should try to make riders have to think their way around, constantly having to make decisions based on how their horse is going. A simple example is if there are two corner fences on a course, one relatively early, the other later, if the horse did not jump the first one very well the rider may be wondering whether to jump the second one.

As a guideline, all regular competition fences should be 5.5m to 5.8m (18ft to 19ft) or so in width except for those which are obviously intended to be narrow or 'skinny' questions. This allows for bad weather and gives riders the

A good wide inviting fence

chance to not all take off and land in the same spot (if they so choose). It is also much kinder to offer wider fences when producing young horses who can often tend to drift one way or the other if they are a little suspicious. For three-day events and park courses the fences tend to be a little wider – more towards 7m to 8m (23ft to 26ft 3in).

When considering fences it is essential to recognize what could go wrong if a horse or rider makes a serious mistake. I am a great believer in the 'what if' theory, for example, what if the horse ends up in the fence? How are we going to get it out? What if the rider is under the horse? How are we going to handle the situation? What are the down sides? We all have to make judgements based on experience and try to imagine every conceivable situation but the reality is that if the degree of risk is unacceptable then the fence should be redesigned or forgotten about altogether.

These considerations are fundamental. We all strive to make courses as safe as possible for horse and rider and to that end we must consider all possibilities and make judgements accordingly. For example, a fence must be built in such a way that should a horse get into the middle of it for whatever reason, it can be taken apart safely and quickly and then reassembled swiftly. Along the same lines, should a horse be stood with its front legs one side of the fence and its hind legs on the other there must be a plan to deal with the situation. If there is no plan then one must question whether the fence remains on a course.

To take a fence apart it must obviously have been put together in an appropriate manner in the first place and this is covered in the next part of this book.

·

Construction

*'The cost of a bad fence is as much
(if not more by the time it has to be
rebuilt) as a good one and therefore it
may as well be right to start with!'*

PRINCIPLES OF FENCE BUILDING AND CONSTRUCTION

Always build with safety as the top priority. It is believed that the size of timber now being used in post and rail fences, oxers, and similar (20cm to 30cm/8in to 12in diameter), encourages horses to jump better with the fences looking more substantial and impressive than say twenty-five years ago. Certainly horses tend to respect the bigger timber and are less inclined to 'flirt' with it.

All fences in or on which a horse may accidentally become trapped or straddled must be able to be easily dismantled in order to extricate a horse without injuring it (or the building team!) and then it must be able to be reassembled very quickly and easily. If there is a concern that there is a better than minimal chance that a horse could get stuck in a particular fence then perhaps it is not the right design for that particular location or, if it is a spread fence, it must have a platform in the top that is strong enough to withstand a horse that may bank it.

Clearly there must be nothing on a fence on which a horse may injure itself, e.g. knots, protruding nails, sharp edges, and there must be no gaps in which a horse may get a leg caught or stuck. Always make sure that any such gaps are either big enough (at least 20cm to 25cm/8in to 10in) or small enough so that it is impossible to get a leg into the fence in the first place.

It is now common practice to rope all rails onto uprights, primarily for safety reasons. The days of wiring (or bolting) rails onto posts have long gone. There is however one exception that can be made although it is not ideal and this is to ensure that there is provision made to cope in a hurry with horses caught up in a fence by ensuring that there are wire cutters readily available.

For roping the most commonly used rope is 8mm white polypropylene which looks good, should last well, is readily available, and inexpensive. The most popular method is one that needs no knots and is self-tightening.

Start by cutting off the required length of rope (which will obviously depend on the size of timber being used); melt the ends of the rope so that they do not fray; stand behind the upright, find the middle of the length of rope and then place it round the top of the post with the ends going towards the approach of the fence (i.e. away from you). Take the rope under the rail to

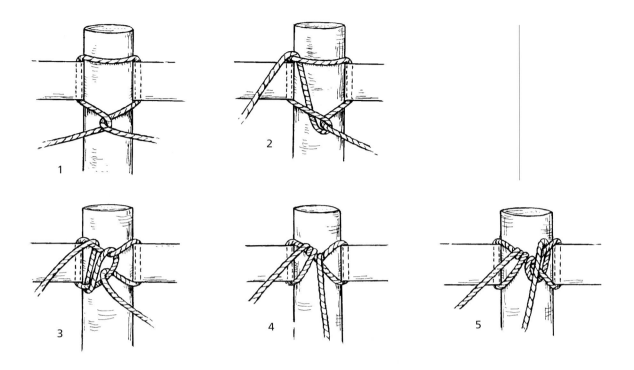

Roping a rail to an upright.

be secured and then cross it at the back of the upright; hold one end of the rope securely and pass the other end up through the gap between the upright and the rail before then pulling it down tight and repeating the exercise; then swap to the other end of the rope and repeat the procedure. It is then just a case of repeating the procedure, alternating each turn from one side to the other until the roping is really tight and the rail fully secured. With practise this process will not take long and it provides an attractive finish which is practical. A useful hint is to not cut the top of the upright off until the roping has been done.

It is essential that there is nothing hidden on which a horse may catch itself through no fault of its own e.g. a back rail that is not visible at the back of a hedge. A horse will jump what it sees, not what it cannot see.

Remember that the cost of a bad fence is as much (if not more by the time it has had to be rebuilt) as a good one and therefore it may as well be right to start with!

COMMON MATERIALS

The following list is by no means exhaustive but will give a useful starting point:

Rope 8mm diameter polypropylene rope although in countries with high levels of sunshine this can perish fairly quickly.

Timber Rails: minimum taper 6m (20ft) lengths for normal usage. Soft wood is significantly easier to work with than hard wood (but it will not last as long) although one must beware of splinters and knots on certain types of timber. Some types of soft wood will deteriorate quicker than others, for instance silver birch will last maybe two years, quite often only one. Larch is generally considered the best looking wood and is also kind on saws and has good uniformity and straightness. Ash often looks good for one, perhaps two years depending on the weather but it splits after eighteen months or so; pine has lots of knots and splinters; beech does not last very long and will start to look unattractive fairly quickly; Norwegian Spruce can last well and look good if it is looked after and has the bark and knots removed.

Wire Regular mild steel (non high tensile) 10 gauge steel wire.

Greenery and tops In most instances everyone is grateful for whatever is available through thinning but Norway spruce is good for both, and Christmas trees are good for dressing.

SIMPLE POST AND RAILS

Ensure that the post is at least 75cm to 80cm (29½in to 31in) into the ground, more if the footing is not that secure e.g. sand. If in doubt add an additional post.

Always step out the lower rail. Vertical fences do not really serve much purpose and certainly do not jump as well. On certain fences, such as a fence at the top of a slight slope going into a coffin, it is well worth considering a third rail on the ground to assist the horse by giving it a take off rail and having less daylight in the fence.

The 'dummy' post (often referred to as a 'godfather') needs to be set 5cm to 6cm (2in to 2¼in) into the ground and should ideally be the same diameter as the rails; on ground that can become very wet it is often useful to put a stone underneath to help prevent it sinking over a period of time although quite often fences will settle anyway and need raising back to height from one year to the next.

MEASURING FENCES

The height of a fence is measured from the point where horses are expected to take off to its highest point, and the drop of a fence is measured from its highest point to the point where horses are expected to land. This means that fences up hill inevitably look very small (and will, by and large, jump extremely easily) and one has to question their value. To get some value from an uphill fence it is often better to move a fence back a stride so that it will look more impressive or certainly to site it in such a position that it can be measured off flat or very slightly rising ground – the precise siting of the fence will be dependent on striding if it is related to another effort/element and the type of terrain.

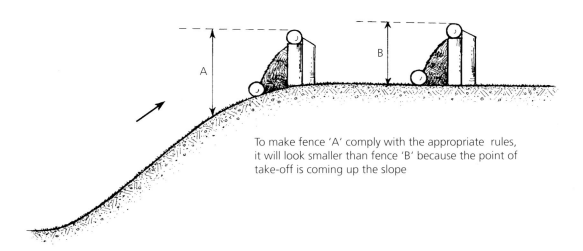

To make fence 'A' comply with the appropriate rules, it will look smaller than fence 'B' because the point of take-off is coming up the slope

The spread of a fence must include everything constructed that the horses are being asked to jump and in the case of an oxer it must include the width of the rails, not just the rails at their highest point. For corners, the spread should be measured where the horses are expected to jump it, and if a fence is designed and constructed in such a way that the intention is that it is jumped at an angle, the width where it is expected to be jumped must conform.

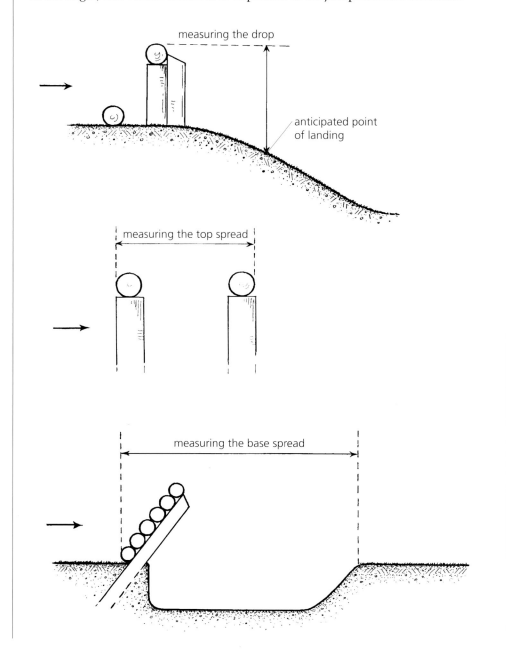

REVETTING, STEPS, AND BANKS

REVETTING

Whatever the type of fence that needs revetting the technique is essentially the same; it is not difficult but it does take time and must be done correctly to avoid having to redo it after a couple of years when it may have started to collapse a little.

The size and shape of the revetting can determine the 'look'; with ditches for example it is possible to have treated half rounds supporting the revetting (provided that they are wired back into the ground), whereas for a big step down it is far better to have the uprights hidden in the drop itself with the sleepers/primary structural timbers bolted onto them, or to build a frame and then fix facing timber to it (this is because horses that are somewhat hesitant can sometimes, although rarely, slide their fetlocks down the face).

Nowadays it is common to use half rounds or unirails (machined half round rails) to face/clad the revetting to either finish the actual construction of the fence or simply for the look of it.

Step clad with unirails for good presentation.

There are some options for the method of revetting. One way is to simply build a wall of sleepers or treated timber and then put a half round on top and clad if wanted; another is to build a frame and then clad it with half rounds.

The finished height and shape will need to be agreed and then start by working out how to achieve these measurements given the materials that will be used. Anything that is going into the ground should be treated, even the small posts that will be wired back to; the reason is obvious.

If building a step or steps using sleepers or similar timbers then start at the bottom and ensure that the level is correct and that it is horizontal otherwise as the face builds up it will end up askew; put at least three treated half rounds or square sawn posts per sleeper 15cm to 20cm (6in to 8in) into the ground or further if necessary (for instance, if revetting a ditch when the bank may be a little soft) so that these uprights are secure. It obviously looks good if the join between sleepers is covered by these. Nail the uprights to the sleeper when they are in the correct position. If the plan is to use two sleepers high for the step or ditch just lay the bottom row at this stage; if the plan is to use three sleepers high for the step or ditch just lay the bottom two rows at this stage. Interlock the sleepers, rather than just placing them one on top of the other, for additional strength just as if building a brick wall.

Opposite each upright it is necessary to then dig a trench at least 1m (39in) (further if the ground is not secure) into the bank approximately 25cm to 30cm (10in to12in) wide. Into this trench it is then necessary to drive in a stake at the same level as and at an angle away from the revetting. Using 8 or 10 gauge soft wire take a length around the post supporting the revetting and the other post in the trench, pull the wire tight and secure it to the upright in the ditch such that it will not come undone/untied. Then twist the wires together using a large nail or similar until they are really tight (be careful not to overdo it so that the sleepers are not pulled into the ditch). Then add the top layer of sleepers and nail the uprights to it.

One more small but important job is to place some 7.5cm (3in) (or similar) diameter plastic flexible drainage pipe inside and at the base of the revetting to avoid any build up of water behind it, and then it will be ready to be back-filled which needs doing carefully so that the wires are not stretched; they will need carefully packing around them by hand before the bulk of the fill is added and the best way to do this is to add 8cm to 10cm (3in to 4in) at a time and pack it tightly before adding the next 8cm to 10cm (3in to 4in).

If the amount of fill is significant it is better to use stone or hardcore until the last 15cm (6in) above which it is best to use 25mm to dust. This will probably settle over a period of time, no matter how good the compaction whilst it is put in, and so there will be the need to top up at some time or other.

The other method of revetting is to create a frame of vertical supports and crossmembers which can then be clad with half rounds. Set treated uprights at the necessary intervals in order to create the required shape – these should not be more than 1.80m apart (just under 6ft); wire these into the bank as previously described and then clad as required. A proper frame will need to be created in order to secure the cladding and this can be done by securing two horizontal pieces of treated timber such as 15cm x 7.5cm (6in x 3in) between the uprights.

Whichever method is preferred is down to personal choice but if the preference is to hide the vertical supports in the bank out of sight it will be necessary to secure the sleepers/facing timber to the uprights using stout coach bolts.

It is also important to remember that all courses are created to withstand the extremes of weather and that if it is wet it is essential that the footing will remain good. Therefore it is advisable to use 10cm to 15cm (4in to 6in) of 25mm to dust to finish off the top of the step/bank before any turf or topsoil with seed is added (preferably the latter so that it helps to bind the materials).

Wiring back into the ground is essential to prolong the use of all revetting.

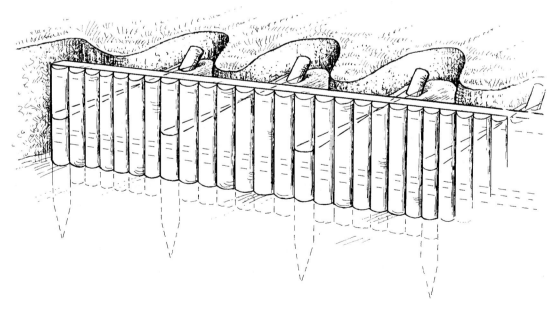

Example of timber frame used in revetting.

The sequence of photographs beginning here and continued on pages 25 and 26 illustrate the construction procedure for this method of revetting.

When building steps it is important to finish off the sides either by round-ing them off using soil and stone or by revetting. The former tends to look better but it all depends on the location of the fence. The advantage of revet-ting the sides is that the wiring can go across the whole bank to help make the bank or step self-supporting. Additionally, when building a big step or drop (over 90cm/35in) it is necessary to have more than one set of retaining wires on each retaining upright back into the bank. The solution is 'if in doubt put more wiring in' – we do not want these to be collapsing! Start off by having some wiring low down the revetting and then have some more higher up.

Once the revetting is completed always rasp or sand off any sharp edges that could possible cause injury; this includes the tops of any/all uprights.

Horses should be helped in their jumping of steps by ensuring that the ground is slightly ramped up in front of each one. Prepare the ground so that over a 2.75m (9ft) distance there is a rise of 10cm to 15cm (4in to 6in). It is an inexpensive way of improving the way steps jump, particularly bigger ones.

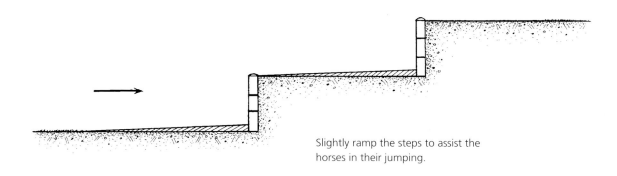

Slightly ramp the steps to assist the horses in their jumping.

Lastly, but very importantly, when designing steps or banks remember that it is essential to have access to all parts in case of an injured horse or, if a horse jumps down one step but will not go down the next, it must be able to get out easily.

BANKS

Often there is the opportunity to use up some spoil and the question is what to do with it. The answer largely depends on what the consistency of the spoil is, for instance, good quality topsoil can be used for preparing take-offs and landings, smoothing out some rough areas, and so on, whereas poor quality spoil has more limited value but can certainly be used in the construction of banks.

Banks can be all different shapes and sizes to cater for all standards and they can be made more interesting with the questions changed by adding fences before, after, and, if room allows, on top.

When building banks the quality of material and the willingness to introduce stone and hardcore will determine how quickly the bank will be usable. To create a bank solely using soil is not a great idea if there is to be significant use and it will need at least two years to settle fully. The concern is that the footing will not be secure or consistent enough to use, especially in wet weather.

If a bank is needed to be used fairly quickly then it needs a lot of hardcore as the foundation (it can be mixed with spoil) topped off with 10cm to 15cm (4in to 6in) depth of 20mm/25mm down to dust with then approx 5cm (2in) topsoil and seed or turf to finish it. The hardcore can just about be anything,

Simple bank with variety of options.

for example building rubble is fine, but what is important is that however the bank is built the layers are compacted as they go in. This will reduce the amount of settlement and will enable the fence to be used more quickly. There is an advantage in mixing some spoil with the hardcore in that it will help retain some moisture in dry times if the top surface is grass. If the top is to be stone dust then this is not an issue.

Depending on the design of the bank it may be necessary to do some revetting. Any step(s) will need the revetting done correctly otherwise it will simply collapse in due course, and as the bank is filled behind the revetting it is essential that this filling is carried out very carefully so that the wires are not stretched. Fill a bit at a time and shovel the stone carefully around the wiring – do not just use a machine to tip on top of it.

Before going to the expense of building a bank it is worth consulting an experienced designer, particularly when planning one incorporating bounce, one, or two stride distances. Guessing can prove a costly mistake.

Steps onto a bank need to be carefully thought out. To build them at maximum height is not a good idea (nor is it necessary); they will be too big and the horses will find them difficult to handle. It is better to be too small than too big. Think about what horses are being asked to do – over a normal fence they have time to get their legs sorted out as they go over the fence whereas up a step they have to be organized differently and the rider needs to adopt a slightly different technique. A useful tip where there are two or three 'biggish' steps is to add a groundline to the bottom one if there is a concern that the horses could perhaps get a little too deep.

GROUND PREPARATION

Another one of the designer's responsibilities, and probably the most important of all, is to ensure that the ground conditions are as good as possible. Competitors' expectations continue to get higher as the value of their horses increases and they will understandably not run on poor ground for fear of injury, and it is fair to say that competitions, events, and schooling facilities will be judged by the quality of their footing – if it is poor word soon gets out and the numbers attending will drop. All too often a large amount of money is spent on the design and construction of a course and the footing is taken for granted, but this is a big mistake and no matter how good a course may be it will count for nothing if the footing is bad. The ground conditions, and their ability to remain consistent throughout, are probably the most important factor to take into consideration when choosing a site for a cross-country course.

One of the attributes of an event horse has to be its ability to run and perform on any type of going in all types of weather. The vagaries of the climate inevitably mean that all will be encountered, wet, dry, hard, soft, hot, cold, but because we have absolutely no control of the weather we must plan well in advance to have the ground as good as we can get it. It is also important to try very hard to have the footing the same for the last competitor as for the first so that it is fair to everyone and that is easier said than done!

Nearly all events experience problems in one way or another with their footing. It could be that stock have to graze the land through the winter, there could be a drought, it could be extremely wet, and so on, but there are certain steps that can be taken to help. Clearly the type of ground and what it is usually used for will determine any necessary action that will need to be taken. It will also determine the time of year that is suitable to use the course, for example there is little point in running on clay either early or late in the season when it may be waterlogged, whereas it is possible to run all year round on sandy, free draining ground even though it may be a little dead on occasions.

The important point in any ground preparation is to think well ahead and know what to do to cover all eventualities. It can take several years to get a good base into a sward and it is well worth consulting a local farmer or agronomist whose knowledge and experience of the ground in good and bad conditions will be invaluable. It is even better if the designer is fortunate enough to see the ground at its worst whether it be high summer or under water when it becomes very obvious where not to go.

With wet ground and courses that get a significant amount of use there may well be a need to consider putting down all weather take-offs and landings to ensure secure footing at the key moments; the down side to this is that it is expensive and commits to certain locations for fences. Undoubtedly there will be quite a lot of maintenance without them and the cost of installation needs to be weighed against the benefit.

A point to remember is that if the take-off needs some stone dust, 25mm down, sand, or similar, to improve it at the last minute it is very important to bring the material far enough from the fence so that there is no way that a horse may pick up too early; normally anything between 4m and 6m (13ft and 20ft) is sufficient and remember to do the practice jump as well so that a) the horses get used to it and b) if the ground on course needs treating you can bet that the practice fence will.

The type of material to put down is important. In wet, deep conditions we need to understand that the footing needs to be made more secure and less deep and therefore material such as 25mm down/Type 1 is good; as it gets pushed into the ground it will also provide a long term benefit. Sand is not good because it does not improve the security of the footing in these circumstances. These conditions are always difficult to manage and if in doubt about the safety of the fence to be jumped, take it out. Horses must be able to take off and land safely without getting held down or slipping on insecure footing.

More significant is hard ground and what to do about it. If nothing else ensure that it is smooth, not rough. Hard, rough ground is the worst. It may mean that it is necessary to hire in some machinery to soften it or break it up, it may mean top dressing to fill rough or poached areas, it may well mean spending quite a bit of money to sort it out and this is an issue that needs to be planned for.

If the ground is hard but smooth the easiest solution for the landings is to put 5cm to 8cm (2in to 3in) of sand down which will take the sting out of the jump for the horses (remember to pick it up afterwards or rake it out); avoid any materials that are slippery and remember to sand the practice fence also. This however does not help with the course itself when it may be necessary to use machinery to break up the ground a little. There are always new pieces of machinery coming out and one of the most popular at the moment is an aerovator which shakes the ground using vibrating knobs on rolling drums. It has the advantage that it does not destroy the top surface, rather it opens it up so that any rain or water goes straight in. Ideally it is then good to get water onto the area that has been machined. Another very good piece of machinery is a 'turf-conditioner' which breaks the ground up a little and then rolls it down afterwards.

It is a real problem when there is really rough ground that has been badly poached (ideally this does not happen because stock management is such that it is avoided!) and then dried out very quickly. The bottom line is that it will be necessary to fill the holes one way or another (top dressing or men with shovels) or, weather permitting, get the rotavator out. The problem with the latter is that if it then rains it will be very slippery but if half the width of the course is rotavated and the other area treated as well as possible all eventualities should be covered. If the answer is to fill the holes it is possible to get away with not doing the whole width of a course by stringing a corridor, thereby reducing the workload considerably. After rotavating it may be necessary to run over the disturbed ground with a roller to break up any lumps or have a stone picking team on hand – it obviously depends on the land but whichever is the case it is imperative to check that no dangerous or sharp objects have been brought to the surface.

Aerovator

Harrows, Dutch harrows, and discs are also very popular for remedial work depending on the ground conditions. Vertidrains are used by a lot of racecourses to penetrate deeper into the ground but are more expensive. If in doubt as to what to do, ask someone experienced in plenty of time – do not wait until the last minute! Consult the national governing body who should give some useful pointers or other organizers, farmers, or indeed racecourses. The key is to have a good working relationship with the farmer/landowner so that many of these issues are avoided in the first place and so that by the time the event is open to competitors and public the course is in its best possible condition with some good grass cover.

Finally, but equally important, it is essential to get the ground back together after a competition. Quite often it is the last thing anyone wants to do but it is undoubtedly time and money well spent and it will save even more time when preparing for the next competition. Sort out the take offs and landings, pull back any material that may have been pushed up to some of the fences by the horses, sprinkle some grass seed and some topsoil if necessary, put the ground back to how you would like it.

A useful tip if fencing off the jumps to keep stock off them is to make sure that the fence is in such a position that when it is removed there is no unintended groundline left behind, for instance, sheep will often walk or lie down next to such a fence and the track that they leave could be interpreted by a horse as a groundline which is not what is wanted.

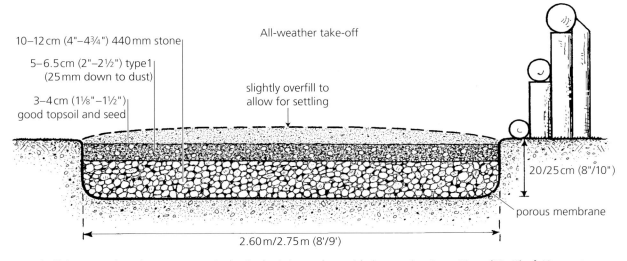

10–12 cm (4"–4¾") 440 mm stone

5–6.5 cm (2"–2½") type1
(25 mm down to dust)

3–4 cm (1⅛"–1½")
good topsoil and seed

All-weather take-off

slightly overfill to
allow for settling

20/25 cm (8"/10")

porous membrane

2.60 m/2.75 m (8'/9')

n.b. if the ground can be very wet or is classic clay it is worth considering putting 8cm–10cm (3"–4") of 40mm stone underneath the membrane

33

·

Design Principles

'Where a fence is sited, and its possible
relationship with other fences, will
determine its severity'

STARTING FROM SCRATCH

One of the best challenges is the opportunity to create a course from scratch where there is no inheriting of another designer's ideas. Sadly these chances do not come along very often and so when they do it is the ideal moment to stamp a personal signature on a course.

There are many factors that designers have to consider before setting foot on a proposed site for a course, let alone coming up with any ideas for fences. Step one is to obtain a copy of the rules that apply to the competition and then to understand them fully. Then it is time to meet with the organizer and landowner so that they understand and accept fully what they are taking on with the consequential requirements. Equally, the designer must respect and understand fully their requirements. Time spent with the landowner and organizer is essential in order to learn as much about the land as possible, what the landowner/host/organizer is looking for in his/her course, where to go, and perhaps more importantly, where not to go! Learn about the year-round conditions so that a building programme can be devised. Also, find out where there may be footpaths or SSSIs (Sites of Special Scientific Interest which are protected) and check whether the landowner wishes to look at fences on his property all year round or whether the fences should be discreetly positioned.

Undoubtedly the most important component of any course is the quality of the footing. This is covered in the section covering ground preparation (see page 30).

Another significant issue is the budget and budget management. A course can cost as much or as little (within reason!) as an organizer wishes but the designer must be able to design to a budget. This is no different to an architect producing plans for a house who, when he is asked to come up with a house design that will cost no more than £75,000 to build, should do just that and not waste everyone's time coming up with a plan for £90,000. The same applies to course design. Seldom, if ever, is there a blank cheque available and I doubt whether a better course would be produced if there was. The key to working on a tight budget is to do less fences but do them properly rather than try to spread the money too thin and end up having to compromise. A bad fence will cost just as much if not more than a good one. Normally, and perfectly understandably, an organizer will wish to have a three year budget and investment plan to build up a course over a period of time rather than go for broke in year one. Designers must be able to assist and work with such a programme.

An understanding of costs of labour and materials is therefore essential as well as an understanding of which are expensive fences to build and which ones cost less. All these costs can and do vary in different parts of the world, for instance, in Australia or the west coast of the USA the timber available is mainly hardwood which takes a lot longer to work with because a) it is so heavy, and b) it is so hard, whereas in Europe we use mainly softwoods that are very easy to work with and significantly lighter in weight. However, hard-woods do last longer than softwoods. There are some fences that are obviously more expensive than others and there are some whose cost-effectiveness is to be questioned, e.g. steps which are more expensive to build than many fences and in some situations can be an extravagant use of jumping efforts.

Discussion needs to take place regarding the layout. Where will the access and egress be for the competitors and spectators? Where will they park? How will they get there? Will they get there if it is really wet? Is the parking in a convenient and user-friendly location? Clearly a course is part of an overall event layout which should be as competitor and spectator friendly as possible. Where will the dressage and show jumping be? Might the cross-country inter-fere with those other phases? If so, can the timetable be adapted accordingly or does the course have to go elsewhere? And so it goes on.

It is a good idea to ask the organizer where he/she sees the event in five years time. What is the goal? How do they intend to get there? Are they look-ing to attract sponsorship and a good crowd? All these issues impact on the cross-country one way or another and it is better to start the project with the longer term in mind even if it appears a little odd to start with – at least this way it will reduce the chance of a major change (with the consequential cost implications) to a layout as the competition grows.

For a one day-event or a hunter trials the start and finish of a course should, unless there are exceptional reasons, be close to the box park so that after a competitor has finished there is not a long way back to the lorry. However, for a three-day event it is not necessarily quite so significant to position the start/finish close to the stables although convenience is still a high priority and it is obviously preferable.

The shape of the course is important for various reasons. The ideal shape for spectators and the course administration is a clover leaf so that all parts of the course are easily accessed with the minimum of walking or driving whereas at the other extreme, a long, thin course is not at all user-friendly; it may achieve a minimum distance but that is all! For television purposes it is important to help with the production costs by allowing a camera to cover several fences if possible, especially the feature fences. The public address contractor will have cabling and sound coverage issues, the emergency teams will want to have speedy access to any problems, and so it goes on. Try to create a focus to the course.

It is no use coming up with a particular route if it is not possible to manage the course. The administration of the course is a major consideration in all planning. There must be good access to all areas for emergency vehicles and officials plus, for the bigger events, plenty of room for spectators and associated facilities such as toilets and food units and their service vehicles.

After all these issues have been discussed it is time to get out on the land and walk everything available At this stage it is very much getting a feel for what there is and the shape of the land. The aim is to come up with a route that is suitable for the competition and satisfies all the criteria previously discussed with the organizer and landowner. Sometimes it is very easy to find a route, other times it takes a lot of working out. There are no particular guidelines about this, it just boils down to gut instinct. It is fairly straightforward to feel whether a course should be right or left handed; the shape of the land and how to put it together will determine this and the best guide is to try to imagine how the proposed route will ride – forget about any fences at this stage. Spend some time alone getting the feel of the land, how you imagine it riding if you were on a horse; think of the picture that you wish to create, ask yourself whether there would be a good flow, whether the horses would be able to settle into a rhythm and then whether your initial thoughts for fence sites will arrive sweetly for the horses. If the answer is 'no' to any of these, think again!

Obviously in thinking about the proposed route a designer will have studied the land for some time and included some of the natural features (if there are some!), including any site(s) for a water fence(s). A word of caution here – if the site has an abundance of features it is very easy to get carried away and use too many of them – remember the flow and do not let enthusiasm get the better of common sense!

Many sites have few natural features and this is where imagination really

Opposite page A good shape for the cross-country will make it much more spectator friendly and easier to manage.

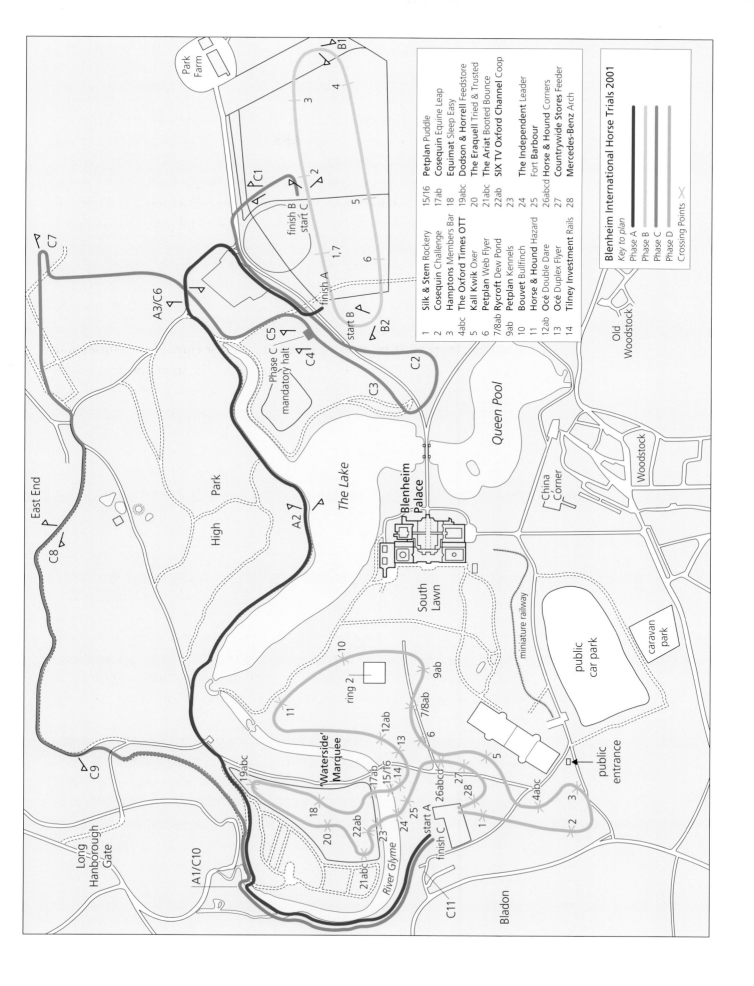

Blenheim International Horse Trials 2001

Key to plan

1	Silk & Stem Rockery
2	Cosequin Challenge
3	Hamptons Members Bar
4abc	The Oxford Times OTT
5	Kall Kwik Oxer
6	Petplan Web Flyer
7/8ab	Rycroft Dew Pond
9ab	Petplan Kennels
10	Bouvet Bullfinch
11	Horse & Hound Hazard
12ab	Océ Double Dare
13	Océ Duplex Flyer
14	Tilney Investment Rails
15/16	Petplan Puddle
17ab	Cosequin Equine Leap
18	Equimat Sleep Easy
19abc	Dodson & Horrell Feedstore
20	The Eraquell Tried & Trusted
21abc	The Ariat Booted Bounce
22ab	SIX TV Oxford Channel Coop
23	The Independent Leader
24	Fort Barbour
25	Horse & Hound Corners
26abcd	Countrywide Stores Feeder
27	Mercedes-Benz Arch
28	

Phase A
Phase B
Phase C
Phase D
Crossing Points

comes to the fore. If the budget is good the opportunity to get more creative exists but this is not always the case. It is perfectly possible to create a good course on flat ground with sensible designing so that there is plenty of variety and the designer uses the ground in such a way that the competitors cannot get going too fast but at the same time they are not twisting and turning all the time. We will look at ways of controlling pace elsewhere in the book but it is a major factor in the planning of the route.

The terrain will also influence the length of the course that a designer should settle on. In all instances there is a window of permitted distances and if the terrain is hilly it is better to aim at the shorter end of the scale whereas one can be bolder with a less demanding site.

Other points to consider at this stage are:

- Avoid starting downhill but if this is not possible, avoid a straight line from the start to the first two or three fences.
- Try to avoid straight lines on courses; use curves to help horses and also to control pace.
- Avoid finishing downhill.
- If you have to climb, climb in the first half of the course before the horses start to tire.
- Avoid running on side gradients if possible; short stretches may be unavoidable.
- Be conscious of camber and avoid adverse ones if at all possible.
- Avoid jinking horses around.
- Keep the groundwork to a minimum to assist with costs; try to work with what nature has provided.

The aim up until this point is purely to find a route that has a good flow to it that will allow for a good design using the natural features. Then come the fences.

At this stage a designer should be building a mental picture of the course(s) and where he/she sees the fences being located. Normally one starts off with too many choices giving far too many jumping efforts which then need to be thinned down to conform to the rules of the competition. This is a good problem to have!

The process of reducing the number of possibilities needs careful thought if the optimum solution is to be found. Getting a feel for the ground by walking it as many times as necessary is essential, thinking about how the possible fences will fit together and imagining how they will ride, working out where the questions will come and how they will relate to each other, how best to use the land, are all issues that are part of the jigsaw.

Once the concepts are planned it is time to get down to detail. At all stages it is necessary to involve and consult with the organizers so that you do not spend a lot of time doing something that they do not like, and particularly now that the course is beginning to take shape. Clearly the proposal has to be budgeted and so the design will need to be documented and costed before anything else happens; this includes any/all groundwork required as a result of the design.

At this point it is necessary to involve the course builder who can assist with the costings. He should give an idea of labour costs and the best places to source materials. As an organizer I always like to get the builder to give a fixed price for the job if possible, breaking each fence down as a cost to build so that it is easy to amend the overall costs if they exceed the desired budget. Understandably course builders are not great fans of this sort of commitment but it is so important to keep to a budget. After all, the quotes can be very specific in what the figures do cover and what they do not cover. Taking this into account, the designer must be very specific and understand that if he/she changes his/her mind there may well be cost implications.

From here on designers have their own way of working with the builder. It is essential that the builder understands exactly what is wanted; try to think of all the possible issues that may crop up, discuss them all until everyone is comfortable, mark the fences out and then get under way. Designers and builders must work closely together, the relationship and mutual respect for each other's craft and skill is paramount. What is important for a builder is that he must not try to second-guess the designer; if something is not quite working then the building on that particular fence should stop until they discuss the problem or the designer next visits. What is ideal is to allow the builder to get the fences offered up and the designer then comes along to set the final heights before the fences are then finished.

In working together there will be discussions on the materials to use, the 'look' of the fences, the style of any unjumpable rails, and, of course, budget management. It is a partnership with success governed by good preparation. Remember to budget for the course preparation pre-competition as well. This can be quite expensive particularly if there are several courses to prepare and dress ahead of a competition. A useful tip is not to have too many brush fences and there are three reasons for this: firstly, brush is not necessarily cheap and can be hard to source, secondly, they need additional preparation work each year and thirdly, stock and deer like to eat them. To counter this, they do look good when done properly and are good fences in many ways – see pages 64 to 65.

Hopefully the end product has a good feel and flow to it, it will be well balanced, and it will be interesting and attractive to the competitors and the spectators.

Horse and rider look confident and relaxed jumping what is quite a big fence into water.

SITING OF FENCES AND USE OF LAND

Where a fence is sited, and its possible relationship with other fences, will determine its severity. The footing, the terrain, the approach, and the landing are all very important, as is how a fence fits into the overall shape and balance of the course. The horses should be allowed to get to any fence in a comfortable way without having to struggle to find a line or make their way over ground that will knock them well off balance; this is unnecessary.

We can also use fences to control pace, either by the type of fence and/or by where it is sited. Judicious use of the land will give a course a good shape. Using curving lines, rather than straight lines, helps the horses. For example, if there is a choice at the start and end of a course as to whether to have fences in a straight or curving line, use the latter; it will make the riders less inclined to come out of the start box flat out and towards the end they will be less able to kick for home concerned about time. If there is no choice, the type of fences to use becomes even more important because we must not be inviting competitors to gallop at our fences with their horses in a poor shape/profile. Granted the riders have the responsibility for themselves and their horses as soon as they leave the start line and how they ride them is up to them but we should at least help where we can.

A straightforward example of how the use of a small area of land can be used to ask very different questions is shown in the diagram on page 44. The use of the ground, the siting of the fence, and the shape of the fence is very important. Whichever fence site is chosen it must link into the rest of the course and it must be suitable for the level of competition.

Whilst considering the siting of fences, it is worth remembering if a course goes from point 'a' to point 'b' around a turn and there is a need for a fence it is preferable to have it sited after the turn, rather than before it, for two reasons. Firstly, it helps control pace, and secondly, it helps the horse and rider set up for the fence. The picture will be much better as a consequence. Clearly there are some fences, such as bounces or gates, that should never be produced on a course in a position where a horse is at speed.

All of us need to be cautious when producing fences on a down slope that we do not look for much spread. It is difficult for a horse to come off its head and jump out and that is why the steeper the slope, the less spread a fence

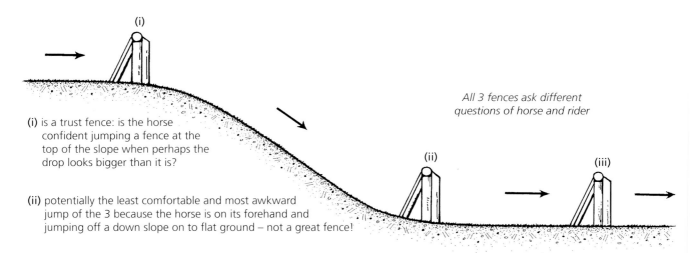

All 3 fences ask different questions of horse and rider

(i) is a trust fence: is the horse confident jumping a fence at the top of the slope when perhaps the drop looks bigger than it is?

(ii) potentially the least comfortable and most awkward jump of the 3 because the horse is on its forehand and jumping off a down slope on to flat ground – not a great fence!

(iii) a simple fence asking the rider to ensure he comes down the slope in a controlled manner: place the fence far enough from the bottom of the slope so that the horses can get off their forehand before being asked to jump: this is particularly important at the lower levels

must have. Again, thinking of how horses jump, make sure that there is no way that the back of a fence can interfere with a horse as it is coming down. For example, the back of a shelter must 'disappear', its pitch at the back should be greater than the front as illustrated on page 72 under 'Unacceptable Fences'.

Drop fences should be sited where the horses are landing on descending ground. This will make life more comfortable for the horses whereas landing on flat ground is much more punishing. If the landing is flat it is worth spending a little money to build it up even just a little bit so that it slopes down which will help take the sting out of it. Avoid landing into rising ground by either re-designing or re-siting the fence or doing some groundwork.

Straightforward uphill fences are, by and large, a waste of time unless placed at the top of a slope beyond the lip. This is because the fence will necessarily be fairly small if it is to measure within the permitted dimensions, according to the rules, from the point of take-off. It is also important to understand that when coming uphill horses cannot readily jump out over spread fences; think about the arc that they will make when jumping the fence and it is apparent that they will be going 'up' more than 'out'.

All fences should give a good feel if ridden correctly and one way of helping the horses in their efforts is to place a fence on ground that is rising very slightly so that they are in a better physical shape and therefore better able to

Sloping landing after a drop

Shelter fence from behind

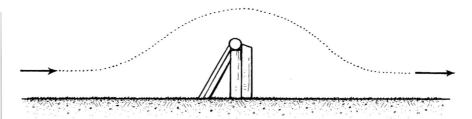

arc of horse jumping fence on flat ground: will land further away from fence

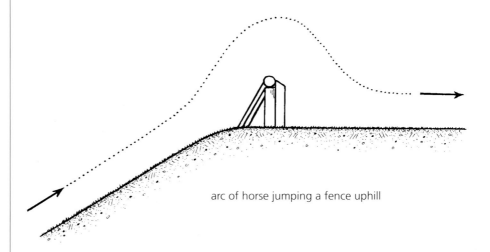

arc of horse jumping a fence uphill

jump. For example, a big fence with a decent ditch in front is much easier for the horse to jump when the approach is slightly rising than when flat, and so sometimes it is money well spent to build up the take off to achieve this.

LIGHT INTO DARK

Many horses will be a little suspicious over these fences since it can often be very difficult for them to see where they are going and what they are being asked to jump. If it is possible always get them into shade before they are asked to jump a fence so that their eyes get a chance to adjust; this is particularly important from bright sunlight to shade which can, as we all know, look very dark. A few simple guidelines:

• Ensure that you know where the sun will be at the time of year when the competition will be held and what shadows it will cause. Adjust the ideas

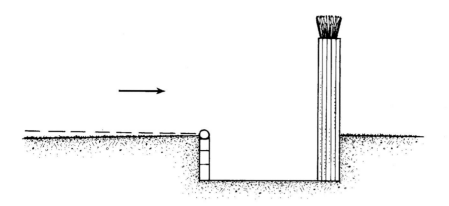

approach is built up to be slightly ascending

The take off has been built up and is rising slightly to assist the horses in their jumping.

for the fence accordingly. Remember that the question should be the same for all competitors if at all possible.

- Use light coloured materials.
- Ensure that the profile is very good; do not have an upright that invites being hit; have a rounded fence or one with a good groundline; the horse must be helped with these fences.
- Be careful how high the fence is; quite often it is better to have the fence below maximum height.

DARK INTO LIGHT

Most horses will jump these fences extremely well; they are keen to get out of the dark and can see where they are going. The areas to pay attention to are:

- Know where the sun will be at the time of year when the competition will be held and ensure that the fence will not be a silhouette. Do not jump straight into the sun.
- Make sure that the fence is very clear to the horses since they may well be paying more attention to what is going on behind the fence than to the fence itself.

BACKGROUND

Be conscious of whether the fence 'disappears' because of its surroundings. Does the fence stand out? Will the horses understand it? An example of this is that in some parts of the world courses are built on very dry land that is burnt off by the sun. The timber used therefore can make the fence blend into its background and as a result it will be necessary to paint or stain the fence a different colour to make it obvious. In much the same way, it is not the best idea to paint fences green in the UK!

Dark into light, this fence is clearly visible.

SHAPE AND PROFILE OF FENCES

Competitors have the opportunity to walk a course several times, make decisions, discuss the options, change their mind, and sometimes to watch the fences jumped before they themselves start, whereas a horse has to evaluate a fence the first time it sees it, whether it is a single element or a combination, in a fraction of a second and then make up its mind what it is going to do. It is essential to remember this.

The shape and profile of all fences must therefore be very obvious to the horse so that there is no way that it will be duped or confused. Fences must have good definition; the top of the fence must be very clear as must the front and back – no tricks. The back of oxers, tables, and other spread fences must always be higher than the front; elements of combinations must be distinguishable either by look, colour, shape, or height (if it looks a muddle to a designer, imagine what a horse will think of it); drop fences should not have much if any spread; upright fences need never be truly vertical.

In expecting a horse to read a question we need to consider what it will see and how it will react. There are many considerations, such as will there be any optical illusions (e.g. hayracks without hay or straw in), will the look of the fence be false and suck the horse into the bottom or allow for misjudgement, will the sun have an influence at a particular time of day casting a long or deep shadow in front of, or on, the fence, what will the background be and can the horse 'pick' the fence? Recognizing these considerations and knowing what to do to overcome them is important – there must be no confusion to the horses. We also need to ask ourselves whether we are asking fair questions, and if the answer is ever 'no' then changes must be made.

Thinking back to how a horse works, if it is surprised when jumping a fence the first thing it does is lower its undercarriage; it wants to get its feet back to the ground where it feels comfortable and safe. Therefore let the horses see what the question is and allow them to decide whether they like it or not. For example, allow them to see through the first element of a coffin fence, particularly if on one stride, if they cannot see over the top of it to the ditch that follows. This is especially important at the lower levels when the trust between horse and rider is yet to be established, but I believe that this is relevant at all levels. At the top end of the sport it is not unreasonable to expect the horse to

These photographs show how the back of the spread is higher than the front giving clear definition to both fences.

50

have the trust in the rider to, for instance, bounce down into a sunken road over a solid upright fence with a good shape, but at the lower levels a kinder fence would be more appropriate and it would be better to have a fence that the horses could see the 'road' through.

With steps or ski jump fences it is always a good idea to have a half round on the top to give definition. Also, if there is a fence such as a step into water it is good to put a 12cm to 15cm (5in to 6in) diameter rail on the top of it so that the horses have to lift their legs rather than just slide down the revetting;

A good example of a fence with a well-presented front groundline.

Horses are able to see through the first part of a sunken road and make decisions based on what they can see.

51

even a 5cm to 8cm (2in to 3in) diameter rail or a half round is better than nothing. This is better because a) it will make the judging easier but more importantly b) it will lessen the chance of a horse hurting itself should it hesitate – we have all seen horses letting themselves down gently rather than making a definite effort, however small that effort may be.

A fence with a ditch in front or underneath will need to have the top edge defined with either a half or full round rail and the ditch must be in contrast to its immediate surroundings so that it is very obvious.

At all levels the fences should look inviting and be 'horse friendly' with the thought that if a rider makes a mistake the horse will have every chance of sorting something on its own.

Half round on top of step down to define the top edge.

below This ditch has good top definition and contrast to its surroundings.

PRESENTATION

Part of the enjoyment and interest in designing fences is how to make them look interesting and attractive. Designers and builders have pride in their work and with little extra effort or cost it is possible to put real character into fences using flowers or by creating a 'theme'. The usual objection to doing this is lack of time or money but at the bigger events in particular it is very important to give spectators, owners, riders, and sponsors something to look at that is attractive and pleasing to the eye, even if it means doing one fence less in order to allow for this.

The difference between a 'naked' fence and a 'dressed' one can be very marked. The most popular, although not necessarily the cheapest, is to put 'tops' in and beside the fence(s); they give a natural look and are easy to put up and take down. Bales are usually cheaper and easier to come by, but quite often, if cost is an issue, the best materials are the offcuts from the building

Well presented novice fence

work itself which can be arranged attractively beside or under fences. It need not cost much but it will give the impression of money having been spent and care taken.

Similarly, a new course has a big advantage over an older one in that the materials will look fresh. For example, a new course with lovely larch rails will not need creosoting in its first year whereas thereafter, once some of the bark has started to be taken off, a coat of creosote makes all the difference. (It is a good idea to take off the bark once it has been broken since it will otherwise hold the wet, thereby shortening the life of the rails.)

It is possible to transform not just the look of a fence with dressing but one can also control which parts of a fence are jumped by placing trees in the jump itself and also where the horses can and cannot approach from by positioning trees/shrubs/rocks/flowers in the appropriate place.

However, none of this will get away from, or disguise, poorly designed or constructed fences. It takes just about as long to build a poor looking fence as it does a good looking one. The finish of the fences, the quality of the joints, the matching of the materials, and the overall presentation all reflect the pride and character of the designer and builder. It is fair to say that many designers/builders put themselves under time pressure; how often do we see work on courses being done well into the evening? Better planning would make life more civilized for all involved and would also allow time for the little touches which put character into a course. Obviously if there is a need to react to weather conditions it is different and if work has to be done late then so be it, but remember, it always takes longer to prepare a course than most of us allow for.

There are various ways of treating and preparing fences and much depends on personal preference, the materials used, and where in the world the course is. Creosote is good where it is permitted and will freshen up the look of a fence. Mix it with red diesel and put on peeled rails or debarked logs and the finish will be very attractive. Stain and varnish are good options but more expensive. Whatever the choice, ensure a matt finish to avoid the risk of any glare and stay away from white painted rails into water which are generally felt to be unadvisable.

For tops and greenery it is usually possible to find some thinning going on not too far away but it may be necessary to buy it in; recommended species for greenery are Norwegian spruce or thuya, and for tops, spruce and Christmas trees.

Usually the style presentation comes down to personal preference and the willingness and pride on the part of the organizer, designer, and builder to present a course well. It is a good opportunity to get imaginative without spending much and it can be very rewarding.

Two well presented fences from the Sydney Olympics giving a good profile and definition to the fences and hopefully making them more encouraging for the horses as well as being attractive to spectators.

CONTROLLING PACE

All designers are, or certainly should be, conscious of how fast a course may ride and if there is concern that competitors may be encouraged to go too quickly there are various acceptable ways of slowing the competitors down. This is a generalization because what is too fast for one competitor is not necessarily too fast for another – it depends on the skill and proficiency of the horse and rider. Having said that, the important point to remember is that whatever is done to try to control pace is done fairly and does not destroy the flow or rhythm of a course.

Usually the time is more difficult to achieve at one-day events than at three-day events because the distances are shorter and the number of permitted jumping efforts means that there is less galloping between fences, but so much depends on the shape of the course and the terrain. Working on the same principle, the reduction in the minimum permitted distances for international three-day events is to be welcomed and will make the time allowed harder to achieve. Undoubtedly working to the shorter permitted distances, whether at one-day or three-day event level, will put a greater premium on jumping the quicker routes and benefit the better riders and horses.

Clearly the type of fence and its location can slow competitors down. A straightforward fence at maximum dimensions will not have much effect whereas a narrow fence or corner requires the rider to set his/her horse up more precisely. The same applies to combinations and related fences.

Stringing/roping of the course, or just a few parts of it, can be used to good effect. If there is a fast section of a course where one wants to slow riders down it is perfectly fair to produce, for example, a narrow faced fence with arms extending back on one, or both sides, and then string each side of the course from just before the fence to just after it.

It is also fair to use the stringing/roping to make a bend or curve which can then be related to a fence provided that it does not take the horses out of rhythm and mess them around. Any stringing/roping used in this situation must always allow a good flow to be maintained.

Another acceptable way of controlling the pace is by swinging the horses back on themselves a little. If this is to be done it must be done in such a way that the horses can stay in their rhythm and it must fit in with the flow of the

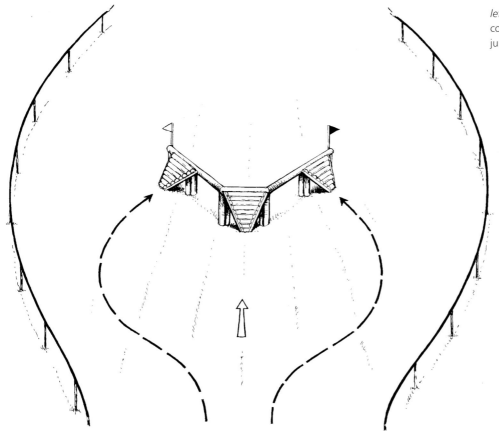

left Using stringing/roping to control the speed/pace if jumping one of the 'arms'.

below Breaking up a straight line to help control speed.

course. It is not something that can often be done more than once or twice in a course, nor should it be.

ARROWHEADS

In recent years as the skill of the horses and riders has gone ever higher we have seen the introduction of what are commonly referred to as 'arrowheads' which require accuracy, control, and good riding in addition to honesty on the part of the horse. It is now accepted that there will be one or two of this type of fence on most courses, including the lower levels, provided that they fit in and are used correctly and in the appropriate location.

It is important to have an ascending face or a rounded shape on such a fence so that it is as inviting as possible; the question being asked is will the horse jump a narrow fence, not will it jump an uninviting nasty skinny little upright.

Looking at narrow fences here are some useful guidelines:

• With a narrow faced fence which is a spread be careful that the spread is not too great, as a guideline not more than 60% of the width of the face.
• At Novice level a face of no less than 1.83m to 2.14m (6ft to 7ft) is recommended.
• At Intermediate level a face of no less than 1.53m to 1.83m (5ft to 6ft) is recommended.
• At Advanced level a face of no less than 1.22m to 1.53m (4ft to 5ft) is recommended.
• Make the shape and profile inviting.

These are for fences with an arm on one or both sides and it is possible in certain situations to go a shade narrower provided the question remains fair for the level of competition. For stand alone narrow fences it may be necessary to help the horses by having some unjumpable railing on one side or the other otherwise the fence could be too narrow standing on its own. It is accepted that horses will find it easier to jump a narrow fence with arms on both sides than a fence with an arm on just one side. Even more difficult is a narrow fence on its own!

An inviting narrow fence with a well-rounded shape and profile.

To see if a rider trusts his/her horse a useful test is to set a narrow fence on a slightly lengthened, but fair, related distance of four or five strides going slightly downhill from a straightforward spread fence; very few riders have the ability to allow their horse to perform this exercise without interfering! As the levels progress so the degree of difficulty can be increased but this is another type of fence that should be introduced early in a horse's education.

A word of caution; it is very easy to overdo the number of narrow/accuracy questions!

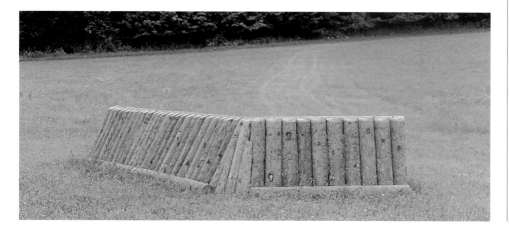

Face-on view of an arrowhead fence with a single arm.

below Narrow fence with brush on each side to help the rider.

59

SAFETY GUIDELINES

As has been stated, safety of horse and rider is, and always must be, the main priority in designing and building fences. Over the years there has been a gradual move to using bigger timber in fences in the belief that the fences jump better, rails are now roped on to uprights rather than wired, and the expertise, skills, and understanding of the fence builders and designers has developed hugely.

But are we, the designers and builders, now producing courses that do not need the horse and rider skills required to jump the courses of even ten years ago? There is a school of thought that feels that fences are now too inviting and do not command enough respect and there is undoubtedly a degree of truth in this.

There are some simple guidelines that have been accepted for certain types of fence as being sensible and practical and are intended to assist designers and builders in their work.

UPRIGHTS

It is generally accepted that true verticals are exceptional or a thing of the past, for instance, very rarely do we see gates on a course without some sort of 'bottom' or groundline, even if just a stone or rock. As always there are exceptions to the rule one of them being a wall, but the key point with upright fences is that the profile is soft to help the horses not get too deep to the fence. If in doubt put a groundline in. This will help the horses, particularly the less experienced ones, jump the fence better. Clearly an upright fence must not be sited where competitors may be travelling at speed.

SPREAD FENCES

The back of the fence must always be a little higher than the front (usually 3cm to 5cm [1in to 2in]) so that the horse can clearly see the profile and shape of what it is being asked to jump. Never ask a horse to jump anything where it cannot see any part of the fence such as a hidden back rail behind a hedge. Consideration should always be given to filling in spread fences. Whether to fill in or not is a difficult question to answer but a useful guide is to fill the top if it is felt that there is a better than average chance that a horse could try to touch down on top or if horse and rider are being asked to think very quickly. If the approach is fine then a straightforward oxer of maximum dimensions should present no problem – after all, it is part of the sport.

below Upright with a good profile with the groundline set in front of the fences. *lower picture* Spread fence with inviting profile; note that the back is higher than the front.

It is worth making a comment on the use of triple bars. If one is to be on a course (and their benefit is questionable since by and large they encourage horses to jump flat and riders to go too fast at them) it is important not to use the maximum permitted base spread; 75% of the permitted maximum is plenty – more than this and the fence will become too 'flat'.

SPREAD FENCES INCORPORATING A DITCH

- Always ensure that there is no risk of a horse injuring itself on any edges.
- Always ensure that there is a way of extricating a horse from a ditch.
- Always have the ends of ditches open or ramped.
- Never have deep narrow ditches.
- Always make a ditch look impressive enough so that there is no chance of a horse being tempted to jump into it.
- If using a ditch behind a fence make sure that a horse will be assisted to clear it; if it cannot see the ditch on approach make sure that the shape of the fence will get the horse over it; if it can see the ditch make the shape of the fence very inviting and always have a rail half buried in the ground to prevent a horse that refuses from slipping into the fence.
- Always ensure that there is a contrast between the colour of the ditch and the surrounding ground; a ditch must not be inadvertently indistinguishable.
- Always define the take off edge of a ditch, even with a small half round, on fences with a ditch at the front; the horse will be focussing on the fence and perhaps not the ditch.
- Do not revet the landing side of a ditch; the exception to this is in a coffin; if it is necessary to support the landing side of a ditch, e.g. in a trakhener, revet to within 25cm to 30cm (10in to 12in) of the top and then round off the ground from the top of the revetting. This way there will be no sharp or hard edges for a horse to land on and risk injury.

TABLES AND HAYRACKS

- Always have the back of the fence 5cm to 8cm (2in to 3in) higher than the front so that the profile is very clear.
- Ensure that there is at least 25cm (10in) vertical 'face' at the top of the fence and, when making the seat for the table ensure that this has a 25cm (10in) 'face'.

- Never have a back seat on a table.
- Always have a solid top in a hayrack some 10cm to 15cm (4in to 6in) below the top to hold the hay/straw. The purpose of this top is to prevent a horse getting into the fence should it try to bank it and it must therefore be constructed of at least 8cm to 10cm (3in to 4in) deep timber with regular cross members. This top will need regularly checking to see that it has not rotted.
- Always make these fences have a good shape at the front so that horses cannot get too deep.

above This shows the 25cm (10in) depth at the top of the fence.

Filled-in top.

BRUSH FENCES

Where there is a solid part as well as the brush part through which horses can pass it is important to have at least 25cm (10in) of brush above the solid structure. This means that the solid part will probably need to be below the maximum permitted height for the particular level. Any brush used must be soft and definitely not liable to cause any injury to a horse. Another important point is to remember that these fences are supposed to be 'brush' and not packed so tight that they are not! It is not an excuse to use the rules in order to create a bigger than permissible fence. Always round off the front top edge of the brush.

Brush Oxers Be aware of the profile of these, the brush at the back must be 5cm to 8cm (2in to 3in) higher than the front and it is recommended to 'platform' between the front and back in case anyone should try to bank it and end up stuck on/in the fence. If a horse is trapped in or on such fences it is extremely difficult to extricate them and so why leave the possibility there?

Platform in a brush fence.

Triple Brushes Always fill in between the rows/elements of brush.

Keyholes The size of the 'hole' must not be less than 1.80m (6ft) in height nor 1.60m (5ft 3in) in width, the top spread must be no more than 50% of the maximum permitted top spread of the particular class, and the solid part of the fence should be less than maximum so that there is at least 25cm (10in) of brush above it. If using a spread always fill the top. There must be at least 50cm (20in) of brush below the solid part of the structure above the fence. Always give them a good apron.

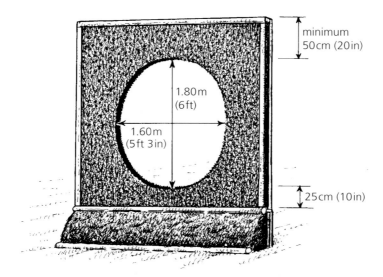

Bullfinches The height needs to be agreed with the appropriate officials but the important point is the thickness. The question being asked is will the horse jump through it and so it must be such that this can happen with ease. Do not make it so thick that it looks solid or too imposing when it becomes unfair. We have all seen the occasional younger/greener horse trying to jump right over the top but they should soon realize that this is unnecessary. It is important to keep a bullfinch as its own question rather than incorporate it with another one because of the variety of ways in which the horses will jump them – some will jump very big, some will jump them as we want them to, and some will drift one way or the other. Only build a bullfinch if it can be maintained properly throughout the day so that it will be the same for the last competitor as the first.

Easy bullfinch which should invite the horses to jump it rather than be intimidated.

FENCES WITH A ROOF

When building these be very conscious of where the sun will be at the time of the competition – we do not want shadows giving false groundlines. There should never be a roof incorporated with a water fence. Any roof should not be less than 2.2m (7ft 2in) from the top of the fence nor less than 3.36m (11ft) from the ground. It is recommended to have the roof higher than these minimum measurements.

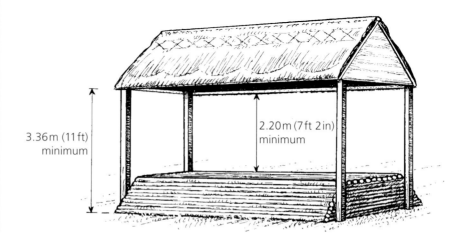

UNJUMPABLE RAIL

It is important that it is very obvious that any unjumpable rail that is erected is clearly not for jumping. Horses, particularly the more experienced ones, will latch on to what they consider to be the fence to jump and make for it and if the unjumpable is not high enough there is a risk of confusion.

Where the unjumpable rail adjoins a fence it is important to recognize that a horse may drift slightly one way or the other and so there are two options:

a) Set the unjumpable away from the fence by 30cm to 35cm (12in to 14in) and put a little brush between it and the fence.

b) Slope the unjumpable away from the fence so that there is nothing on which a rider can get caught and flipped off.

below Unjumpable rail joining two parts of the same fence.

Unjumpable rail sloping away from the fence.

GENERAL POINTS

Brush aprons When building these ensure that the apron is solid/strong enough not to collapse if a horse puts a foot on/in it. If bales are used underneath greenery/brush for instance, make sure that the bales are left whole. Also ensure that it is not possible for a horse that puts a foot on/in the apron to get tangled up in any wire or string that may be used to hold the greenery/brush in place. The wire must be well covered by the greenery/brush.

Ensure also that there are no gaps in which a horse can get its leg through between the apron and the top rail. If in doubt use a half round or similar (straw that is packed really tightly can also work) to prevent this possibility. Always ensure that the top rail is very visible and brief the fence judge to make sure it stays so throughout the day – it is very easy for the apron to get fluffed up and hide the rail which is clearly not what is wanted. As with all fences it is essential to make sure that the apron does not make the fence too flat.

Take-off rails It is important that there is no way that a horse can accidentally get a foot underneath a rail at the base of a fence and therefore always sink the rail into the ground by at least 10cm to 15cm (4in to 6in) and secure it in position. This is particularly important if the rail is used as a support to a shallow scoop or ditch because, apart from the safety factor, we do not want the edge of the scoop or ditch to break down.

Ascending rails and palisades If building a fence using rails or half rounds across the fence horizontally ensure that the bottom rail is dug in up to half of its width. With palissading always ensure that the materials used will withstand a horse stopping and banging into it and also that it is into the ground by at least 10cm (4in).

Gaps between rails In order to reduce the risk of a horse getting a leg caught between rails it is important that any gap between two rails should be less than 7.5cm (3in) or greater than 20cm (8in).

Securing lower rails There is a very good case for roping (not too tightly) just one end of a lower rail so that should a horse put a leg in the gap for whatever reason the rail can move back if pressure is applied and then be easily put back in position.

The top rail can be seen clearly in the green apron.

below Take-off rail sunk into the ground.

Sloping rail fence, note that the bottom rail is half in the ground.

below Sloping rail fence showing the maximum size gaps between rails.

Sharks teeth Never have the 'teeth' adjoined at the top or bottom; always leave a gap of 10cm to 15cm (4in to 6in).

above Always leave a gap between the 'teeth'.

Cross question Put a tree in the middle of the diamond to deter horses from getting into the fence; if in doubt, platform the diamond provided there is no chance of a horse getting straddled on the fence – if the platform is constructed it will make it very difficult to take the fence down in order to extricate a horse. If in doubt, avoid this type of fence, and do not use it at the lower levels because it may confuse the less experienced horses.

right and below Put a tree in the middle of each 'diamond' to keep horses jumping the fence where it is intended to be jumped.

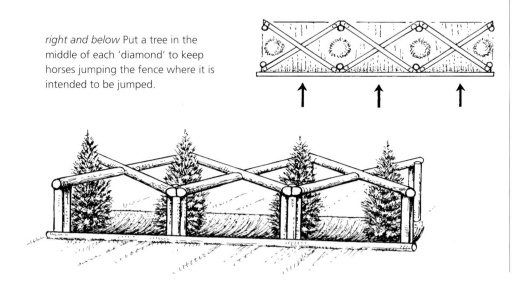

UNACCEPTABLE FENCES

Clearly we must not produce fences that confuse horses, nor must we half-stride them (build combinations or related fences that are on incorrect distances), but there are also several fences that are not part of cross-country and some that are totally unacceptable because they ask unfair questions and punish a horse. Most of the fences that follow are discarded because of what they do to horses physically; it all comes back to the mechanics of the horse.

Top of the list is a drop off a step down to then immediately bounce over another fence. Think what it does to a horse physically, ask yourself whether this is acceptable, and then discard the idea for ever! Similarly double-bounces do not achieve much. Nearly all designers try them sometime and most then discard them as a gymnastic exercise best left for training purposes.

Open oxers with a drop are unacceptable because the horse's natural desire, especially at the lower levels, is to get its feet back on the ground and the last thing we want is the horse to put down in the fence. They are more inclined to 'hug' this type of fence and not get up in the air as much as usual. Big spread fences with drops are a bad principle for much the same reason although shelter shape fences are acceptable because of their shape and provided the width is minimal and the angle of the back of the fence is such that there is no way that a horse will catch it on the way down as it is jumping.

Should we have spread fences into water? Nothing more than sloping rails, a coop shape fence with little spread, or a log or log-shape fence should be on a course for the same reason as before – too many incompatible questions are

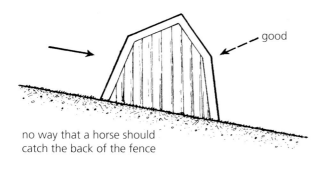

good

no way that a horse should catch the back of the fence

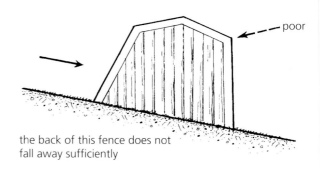

poor

the back of this fence does not fall away sufficiently

being combined if a horse is asked to jump a wide fence into water. Again, the natural instinct for the horse is to want to get its feet down and not jump extravagantly over the fence. With a spread into water there is every likelihood that the horse will catch the back of the fence and this, sooner or later, will stop the horse jumping this type of fence – not what we want.

Other examples of fences to avoid are:

- A bounce into a coffin.
- An uphill fence with too much spread lest the horse catches the back of it on the way down. Imagine a horse approaching up a slope – it can jump high but it cannot travel forward enough to carry the spread as well.
- Single steps less than 45cm (18in) high are more of a hazard than a jump; fine for schooling young horses at home but not great competition fences.
- Half coffin with the ditch first – the horses will be focussing on the second element and may not notice the ditch until very late.
- All flat fences that are ill-proportioned.

Brush fences where we are inviting horses to go through the top of the fence must not have a hidden rail at the back; any retaining half-round must be lower than the one at the front. The same applies to steeplechase fences.

Jumping into rising ground is another one to avoid because of what it does to the horse's legs and joints. It can be very jarring and uncomfortable. The obvious exception to this is in coffins and occasionally it is acceptable if a horse is travelling uphill jumping a fence uphill when his body movement is in an upward rather than forward direction.

All of these are examples of fences that punish horses unnecessarily and there will be more along the same lines. The hope is that we recognize the bad fences before they are ever jumped and then avoid them.

THE EARLY AND LATE FENCES

Quite often these are the most time consuming fences to settle on, particularly the last few because of their importance which is often underestimated.

With the early fences the principle is to provide the riders with some jumps that they can use to give their horses and themselves a chance to settle any nerves and get into a rhythm, i.e. to act as confidence builders, whilst at the same time getting them up in the air. These fences should not necessarily be small. The first fence, and possibly the second, will, at the lower levels particularly, perhaps be below maximum height and from thereon there is no reason why the fences should not be up to height. Small fences do not do the horses any favours. What is more important is that these fences are inviting, straightforward, and encouraging, giving the horses the opportunity to get into the course before they come across the more difficult ones. It is important to get the horses jumping and paying attention so that they are not suddenly surprised when faced with a question.

At the other end of the course the last two or three fences are in many ways some of the most important ones, particularly at three-day events, because the horses will be starting to tire and there is a need to keep them interested and jumping. Common sense tells us that we should be avoiding fences that are asking too many questions late in the course and that may inadvertently cause injury as muscles begin to tire. It is sensible therefore to avoid drop fences or turning questions at this stage of proceedings. Having said that we must keep the horses jumping and respecting what they are faced with. The fences should be up to height, present enough of a question, and be interesting. Certainly they should not be small which may cause horses to become a little casual. Also, when planning the route, avoid running downhill at the end of the course if at all possible.

Another consideration is to keep control of the pace; the last few fences must not encourage riders to go faster than is safe. For this reason straight lines should be avoided unless there is absolutely no alternative; try to use corners/bends for the reasons previously described and do not give competitors the opportunity to get up too much speed. Avoid steeplechase or flat fences or fences that do not encourage riders to respect them; keep the horses jumping and the riders paying attention.

Good example of a late fence – big but straightforward.

Lastly, the run in from the last fence. When planning where to have the finish relative to the last fence it is preferable to work to approximately 50m (55yds) or so. Any longer and we see competitors riding a racing finish which is not particularly attractive nor is it very good for the horses. Under international rules there is a requirement at three-day events to have this distance between 35m (38yds) and 70m (76yds). The run in itself should be 120m (130yds)or so if at all possible, particularly at the higher levels when the horses will be travelling faster, and preferably slightly uphill which will assist them in pulling up.

USEFUL TIPS

- Very few horses jump ditches, corners, drops, or oxers straight; most drift one way or the other in the air.
- Distances on three, four or five strides are often more difficult for riders than one or two strides; there is always the temptation to take a pull or over-ride!

- Riding straight is generally more difficult than riding on a turn. Three or four strides in a straight line to a narrow fence can often be more difficult than three or four strides on a turn to a narrow fence.
- If the terrain allows, think about the riders ending up a little behind on the clock after the first quarter or third of a course so that they have to work out how and where to make the time up (even if it is possible).
- If designing for a three-day event do not have the first or second cross-country fences as green or brush fences; remember the horses have shortly before completed the steeplechase phase, jumping at greater speed and flatter, and we must now get them up in the air.
- If a course jumps easily one year avoid the temptation to rush to make it more difficult. If the course is of the right standard it means that the competitors are of the appropriate standard. However, if it is too easy for the level then it should be 'upped' a little but address this carefully.
- If the plan is to increase the level of difficulty on a course remember to look at the overall picture, do not simply change one or two fences to make them more difficult; this merely leads to an unbalanced course.
- It is good to err on the side of caution; better to be too soft than too demanding.
- If in doubt about the safety of a fence once it has been built take it out and start again, even if it has cost a lot of money.
- Do not half-stride horses.
- Any fence in which a horse may become caught must be built in such a way that it can be easily and safely dismantled and then quickly rebuilt.
- Fences to be jumped uphill generally jump better those on a down slope.
- Avoid fences that 'suck' horses into the bottom.
- Avoid false groundlines.
- Avoid optical illusions, for instance, always fill the insides of hayracks with hay or straw.
- If in doubt, ask – don't just hope!
- Everyone should be working together – designers, riders, officials, builders – we're all on the same side.
- Remove the bark off rails after it starts to come away to prolong their life; always check for knots and remove them.
- If packing a hedge with greenery to make it suitable for competition purposes always remove the greenery afterwards to let the hedge continue to grow.

- All groundlines and take-off rails must be firmly secured.
- Be careful that a fence does not blend into the background – make it obvious and always ensure that there is a good contrast between the fence and its surroundings. This also applies to groundlines – ensure that the groundline will always stand out, do not have it the same colour as the ground or footing material.
- If faced with a ditch that is too narrow and/or deep to be usable one solution is to pipe the ditch and build over it.
- Always ensure that the tops of all posts are rounded off and smoothed off; there must be no rough or sharp edges.
- For fences with a drop always take off the tops of the uprights at a steeper angle than normal; horses tend to hug these fences and will come down earlier, hence the need to remove any protrusion or hazard.
- Be conscious of using narrow 'U' shaped trees as jumps; horses often need help to jump these by ensuring that there are good wings to hold them in to the fence. A horse does not reckon to jump through a tree!
- Always use tall flags, not less than 2m (6½ft) high; this will make the judging of certain fences such as corners much easier when it has to be decided if a horse's head has passed inside it.
- Use plastic ties to secure flags that may be knocked down if a horse is attempting to run out or is jumping close to the flag, e.g. on the apex of corners and on narrow fences.

·

PART FOUR

The Fences

'Understand the consequences of what you are designing.'

OPTIONS AND ALTERNATIVES

In the simplest form an option is a choice between two lines or routes that may be asking different questions that do not entail wasting time whereas an alternative is generally an easier fence on a route that is time consuming.

The difficulty in producing a fence with options is to ensure that there is an even balance and degree of difficulty between them. A designer will try to offer options so that the rider can choose which particular route or line suits his/her horse. A simple example would be two logs, one sited at the top of a slope, the other half way down the same slope.

Using this same example, an alternative to the log at the top of the slope would be, for instance, a log at the bottom of the slope at a different angle requiring a different line altogether to be taken that would take much more time to jump.

WHEN TO USE THEM?

Options are easy as a design decision; getting the balance right is the difficult part so that the whole fence is used and some of it is not, therefore, a waste of money. Alternatives, particularly as the levels become higher, are a different matter in that, at some stage, the designer has to ask himself/herself whether a fence needs an alternative or whether it is perfectly reasonable to expect competitors and horses to answer the question being asked. This is particularly relevant at the higher level one-day and three-day events when the significance of qualifications is much greater.

Clearly it would be wrong to overdo the number of alternatives so that it is possible for a horse or rider to attain a qualification having only jumped easier alternatives all the way around a course – it is essential that competitors have to jump fences of the appropriate standard and are not able to totally circumvent the question(s) being asked. It is fair to say we all are guilty of overdoing alternatives at some time or other.

Another very important issue that comes into the equation is the time factor. For instance, at three-day events competitors normally expect to be able to take one long route on a course and still make time but as designers we do not want someone to take the long routes all the way and still get there. This is not that easy if one wants to give a good flowing course since there will inevitably be good opportunities to make up time.

At the lower levels where the emphasis is on education time is not such a big issue and if 30% to 40% of those who start make the time it is not a disaster provided the course has ridden well and the number of alternatives has not been overdone. Such a statistic would not be very satisfactory at the higher levels when ideally few competitors make the time.

One point worth noting is that one can never justify having a fence that is too strong for a standard by accepting it because there is an alternative. All too often the comment is made 'That's ok because there is a long route' – it is not! If one is building or designing a Novice course, make sure that it is just that, with all the fences, and any options and alternatives, at that particular level.

Along the same lines, an Intermediate course is not a Novice course with a loop of bigger fences added on to achieve the required distance, nor is it a series of Novice fences that are simply raised or enlarged a little – there is a difference in technicality and skill level required as well.

Another instance for using an alternative is to allow the competitors the opportunity to get back into a course after a refusal. In certain situations it may not be possible for a rider to represent his/her horse satisfactorily at the element they have refused at and so the need arises for another solution.

UNJUMPABLE RAILS

It has been usual to have to join the alternative elements of the same obstacle with unjumpable rail. Now in most nations it is possible to avoid having to do this by flagging the quick route and the option/alternative separately with the same number/letter, thereby saving on labour and materials whilst at the same time giving the designers a greater degree of flexibility and creativity in their work. This also gives better viewing for spectators and television at the bigger events.

A word of caution, however. In removing any existing unjumpable rails the fence in question could be changed quite dramatically, particularly if it is a narrow one. The unjumpable rails on narrow fences effectively provide a wing and consequently the fence is much easier.

right Examples of options at the last ('D') element of this Sunken Road where there is the choice of some rails on a bounce distance or a triple brush on a long one stride.

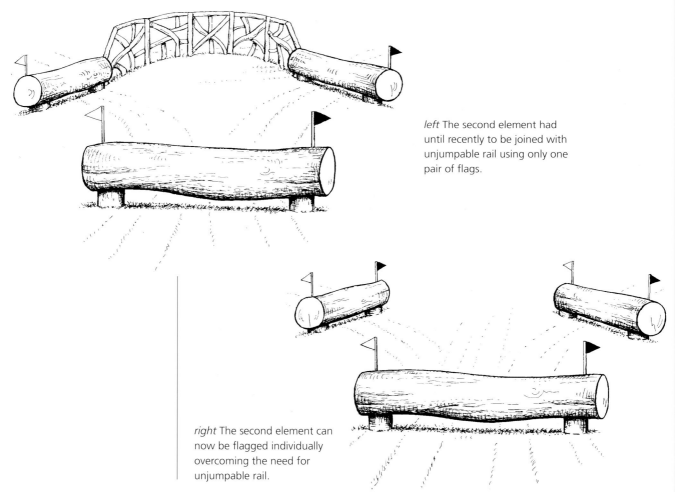

left The second element had until recently to be joined with unjumpable rail using only one pair of flags.

right The second element can now be flagged individually overcoming the need for unjumpable rail.

COMBINATIONS AND BOUNCES

COMBINATIONS

A good combination, no matter what the level of competition, should give horse and rider a good feel if ridden correctly and should hopefully not punish a horse if a rider makes a mistake. The distances should be fair and reflect what the designer is looking for.

As with all fences the questions a designer needs to ask himself/herself are the same – what is the purpose of a combination? What added value does it bring? Why do we need them? What are the benefits? How and where should a combination be introduced into a course? How many should there be? The answers to these questions will vary depending on the level of competition.

The purpose is to check that the horse is confident, quick thinking, and athletic, and that the rider has the ability to present his/her horse in the correct shape at the correct pace to enable it to jump the fence well. Combinations serve a variety of purposes in addition to these ones; they are part of a course, they control pace, they make riders have to think and make judgements based on how their horses are going, they provide interest for the horses, riders, and spectators. When to introduce them depends on the overall character of the course but certainly the aim is to give everyone the chance to settle into a course before asking a question, however simple, and therefore one would not have a combination before fence four or five. Similarly, at the end of a course one is looking to finish the horses off on a good note and so, given that a combination is more of a test than a straightforward single effort it is appropriate to not have one within the last two or three fences if at all possible.

At all levels it is reasonable to expect a horse to go on a regular cross-country stride distance (see table of distances on page 164) and these distances should not be messed around with. As the levels get higher so the scope for adjusting the distances becomes available, particularly when expecting horses to be able to shorten and still stay neat, but this scope for adjustment is not much and certainly there is a direct relationship between the distances in a combination and the actual site and size of the fence(s). Whatever, one thing not to do is to mix and match one's expectations, i.e. do not have the first half of a combination on a long stride and then have the next stride as a short one; keep the distances fair and related and such that they will not punish a free flowing horse (this does not mean one that is going too fast!).

The degree of difficulty varies with the experience of horse and rider. At the lower levels all combinations should be simple and straightforward with an easy approach bearing in mind that we are looking to give horses and riders confidence and experience. Horses at this level are by and large in the early stages of their careers and we should not expect too much of them. It is quite kind to offer angled elements in combinations so that riders can choose the line that suits their particular mount. As the levels progress so the opportunity to use more interesting terrain presents itself, the degree of technical difficulty will increase and the types of questions asked can become more sophisticated. No matter what the level there are some basic principles that should be adhered to:

• Know what you want to ask, why, and how to achieve it in a positive way.
• Never try to catch a horse out; riders can walk the course as many times as they want but the horse has to assess the fence in a fraction of a second and then respond to what it sees.

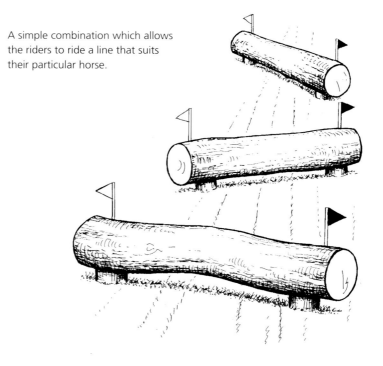

A simple combination which allows the riders to ride a line that suits their particular horse.

- Make the question very obvious to the horses so that they can make a decision.
- Do not try to get too clever.
- Understand the consequences of what you are designing.
- If when it is built it is no good (and we have all done this!) pull it out and start again.
- Always bear in mind that there must be a way of a horse being able to be re-presented after a refusal at any element.
- Always ensure that there is a way for a horse to get out of a complex if it refuses at an element.
- Be conscious of aesthetics; often it costs no more to get creative with the look of a fence than it does to be simple.
- Do not be tempted to add another element if a fence rides perhaps too easily unless you are aware of the impact on the particular combination and on the course overall. We have all seen this happen and often it produces a worse fence as a consequence with a bad balance and flow.

A useful tip is to lay any proposed combination out on the ground using some light rails before starting construction so that modifications to the design can be made before incurring any expenditure.

BOUNCES

Bounce fences are essentially gymnastic exercises. They are made more interesting by their look and location and demonstrate the horse's athleticism and confidence and the rider's ability to present their horse correctly at a fence. However, it is essential that these fences are located in appropriate situations with the appropriate profiles whereby the horses will be given every opportunity to understand and then jump them correctly.

They should not be situated at the end of a gallop where there is a possibility of riders coming too fast to them. They should be sited in such a way that the rider can organize his/her horse and where the pace will not be too quick – off a turn is normally a good spot or related to another fence say four or five strides before on a slightly curving line i.e. where the horse can get his hocks underneath him and any pace has been taken off. Help the horse regardless of the rider.

As far as distances are concerned, for straightforward bounces, at the lower levels use 4.4m to 4.5m (14ft 6in to 14ft 9in) and for the higher levels use

4.55m (15ft). These are measured apex to apex regardless of the style of fence; disregard the distance between each element – the back of the first element and the front of the second element are irrelevant as far as the arc of the horse's jump is concerned.

There are some recommendations as follows:

- Ensure that the elements of the bounce do not blend into each other so that there is little chance of a horse trying to jump it in one.
- Ensure that the horse will be able to understand the question.

- Never ask a horse to jump the first element without being able to see and read the second one.
- Make the profile of the fence inviting; never have any element as a true vertical.

Bounces with bigger spreads (not including logs or simple fences with aprons) are sophisticated fences and therefore only appropriate for the higher levels.

For bounces into water (not really suitable until Intermediate or CCI**/CCI*** level) the distance will be a little less, more towards 3.96m to 4.12m (13ft to 13ft 6in), the reason being that we do not want to have the horses standing off the second element and jumping too flat and long over a fence with a drop; the horses will need to be closer to the second element into water than they would be for the second element of a regular bounce so that they are not having to reach for it. Everyone has their own way of setting these fences up on a correct distance but one way which is to be recommended is to offer up the second element and then stand at the point at which you want the horse to take off; obviously this is also the landing point for the horse over the first element and so from this it is possible to work out where the first element

Bounce into water, both elements have good shape.

should be. After setting the positions of the elements it is then time to set their heights. A word of advice – unless you are absolutely certain when setting up a bounce into water don't do it, too much can go wrong with a bad fence in this situation!

The other vital issue to have in mind is the shape and profile of this type of fence. The horses will obviously be looking to drop and are inclined to snuggle over the fence quite often giving it a bit of a rub. A rounded shape is therefore essential and all materials must be smooth or soft.

A quick word about double bounces which very occasionally appear on a course. Most designers try this once (at a higher level) and then discard it as a poor fence. It is generally considered to be solely a gymnastic exercise best left in the training field, not really part of cross-country riding. Very rarely do these fences jump consistently well and the strong recommendation is to forget about using them.

Downhill bounces are not good fences and bounces of oxers/parallels should not be permitted – again, too much can go wrong.

COFFIN FENCES

These are classic event fences and the sooner horses are introduced to them the better. I would advocate their use, if appropriately sited and fitting in as part of a course, from the lowest levels so that horses learn about them and accept them as part of everyday life.

The severity and design at all levels is critical. At the lower end of the sport they should be very simple and very inviting. A simple log on a flat ground approach, two or three strides to a small ditch or scoop no more than 60cm (2ft) wide and 60cm (2ft) deep, followed by another two or three strides to another small log will be a very good introduction to this type of fence. The logs should be no less than 4.5m to 4.8m (14ft 9in to 15ft 9in) wide (although up to 5.5m [18ft] would be preferable) to allow a horse to 'travel' a little one way or the other and the edges of the ditch should be showing by 5cm to 7.5cm (2in to 3in) above the ground on the approach side. Use a half round to achieve this thereby avoiding sharp edges and ensure that any edges on the landing side are rounded off or have a half round at ground level.

As with bounce fences this type of fence should not be sited at the end of a long gallop. A horse must be able to understand the question immediately and this probably means that it will be able to see through, or over in good time, the first element, certainly at the lower levels. This depends on the approach, the distance before the ditch, the degree of slope on landing, and the size of the horse.

A useful guide is to walk to the fence a few times whilst setting the heights and make a judgement. The question is of the rider's ability to ride and the horse's confidence; can the rider present the horse at the right speed in the right shape and in the right place to jump the first element? If this happens, the rest of the fence should flow well. It is also important to allow the riders to ride positively to the last element, particularly if going uphill, allowing the horses to move on a forward stride rather than on a holding one; they should be able to 'power' out over the last element and not have to be strangled!

The heights of the first and last elements are extremely important, particularly the former. Normally coming out it is possible to have the fence at or just below maximum height. The crucial element is the first one where 2cm or 3cm (¾in or 1in) can make a huge difference in look and severity. Rarely is the first element at maximum height, it does not need to be. Keep walking the line on the approach to assess where it needs to be; work out when the ditch becomes apparent to the horse and ask yourself if you are happy; if not, lower the first

First element of a coffin that horses can see through to understand the question.

element bit by bit until you are (do be conscious that the horse's head will be higher than your eye).

The distances between the elements in these fences will vary a little depending on the approach and the severity of the slopes down and up. It must be understood that once a slope reaches a certain severity a horse's stride will shorten when going downhill and this must be taken into account when building these fences. Regular distances vary between 5.5m to 5.8m (18ft to 19ft) and 6.7m (22ft) but these must be taken on a case by case scenario.

The ditch itself should not be too wide or too deep (or indeed too narrow) but it must be obvious and clearly defined. Anything less than 76cm (2ft 6in) inside measurement at the lowest levels should be avoided. Depth is another issue: a minimum of 69cm (2ft 3in) is recommended provided the ditch is not then too narrow and deep; for British Eventing Novice horse trials a useful guide is 1.22m (4ft) inside measurement with 76cm (2ft 6in) depth going up to 1.52m to 1.6m (5ft to 5ft 3in) inside measurement at the top levels with 92cm to 1m (3ft to 3ft 3in) depth. One always must be conscious of the fact that a horse may get into the ditch and there must be a means of getting them out. They must be imposing enough to the horses yet shallow enough to be man-

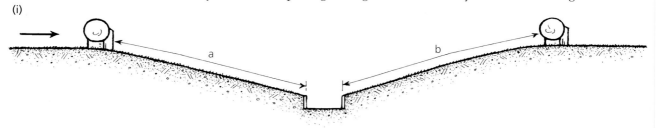

comparison between shape of ground and distances

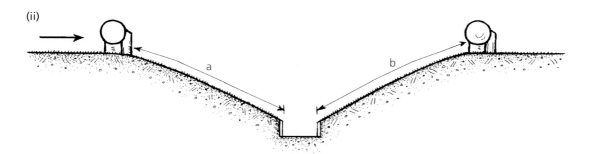

In diagram (ii) the distances 'a' and 'b' will be slightly shorter than in (i) because of the steeper gradient which means that the horse will not be moving as much in a 'forward' direction

ageable if a horse gets into one. It is therefore a good idea to have the ends open so that horses can walk out if necessary. Ditches with the ends filled are to be avoided if at all possible but if for structural reasons there needs to be some sort of revetting then a stone ramp with dust or sand on top must be put in at each end to allow horses to walk out in the unlikely event of getting in there in the first place.

There should just about always be an alternative to the first element which probably then means that there will need to be an alternative to the final element. Such an alternative must not be an 'elbow' on the take-off side of the first element; this is for safety reasons lest someone who has a refusal at the first element is tempted to try to jump the alternative from a standstill. A way to overcome this is to make the alternative a little 'remote' from the quick route and to then number the fence with two numbers or to make the alterna-

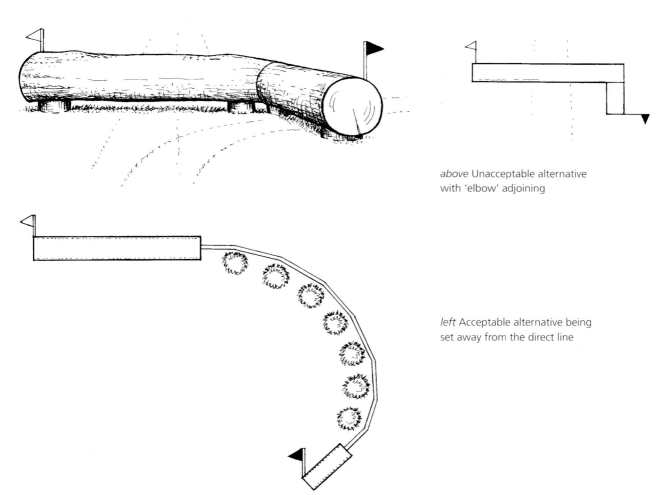

above Unacceptable alternative with 'elbow' adjoining

left Acceptable alternative being set away from the direct line

tive on the back of the quick route. Remember that all alternatives must flow if at all possible – think of the picture that we want to see, it is not of a horse being hauled about by the rider trying to make impossible or uncomfortable turns.

I do not believe that a bounce distance in this type of fence is appropriate. To bounce in or out? Why? What does it add? What purpose does it serve? What does this do to the horse? Whatever the answer there must be better ways of asking the question.

Sometimes there are half coffins on a course with the ditch as the first element. This is not a good idea because the horse generally focuses on what would then be the second element as it approaches and is somewhat surprised when it sees a ditch unexpectedly. For this reason they often do not jump well and are not recommended. In much the same way a half coffin with the ditch as the second element is somehow an incomplete question; the usual last element tends to finish the horses well.

Always make these fences inviting to look at regardless of the level. We all want to encourage and assist the horses to jump the fence in a comfortable way. The first element is really what a coffin is about; the horse will normally make a decision whether to jump when it sees the ditch and once over the first element things should be simpler for the rider (unless of course the last element is an accuracy question). Remember that when a horse is surprised it will drop its legs, and so avoid surprises if possible and keep the profile of the elements soft and obvious!

Quite often it is sensible to have a third rail, particularly on the first element, to help the horse assess his take-off point and prevent him getting too deep. It is particularly recommended if the slope down to the ditch is quite steep or if there is a lot of daylight in the fence. This third rail should always be on the ground and stepped out as per the principle of the second rail.

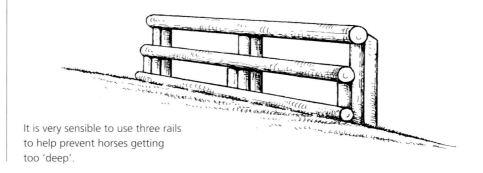

It is very sensible to use three rails to help prevent horses getting too 'deep'.

SUNKEN ROADS

Many of the principles that apply to combinations and coffins apply to these fences also. The question is not dissimilar to a coffin fence; the question is one of confidence, athleticism, and honesty combined with the rider's ability to set his/her horse up correctly.

For a designer a Sunken Road gives a wonderful opportunity to get creative with shapes, distances, and 'look'. The variety of these characteristics is almost limitless but, again as with coffins, the trick is not to make them too difficult by making the first element in particular too severe or big. For the lower levels it is sufficient to make them very straightforward so that they are a good introduction, for example, two strides between the first element and the step down, two strides in the bottom, and then one or two strides to the final element. As the levels progress so the difficulty of the questions asked can be increased so that at the top level it is perfectly fair to bounce in, stride in the bottom, and bounce out (with an alternative long route).

Classic Sunken Road with time-consuming long route.

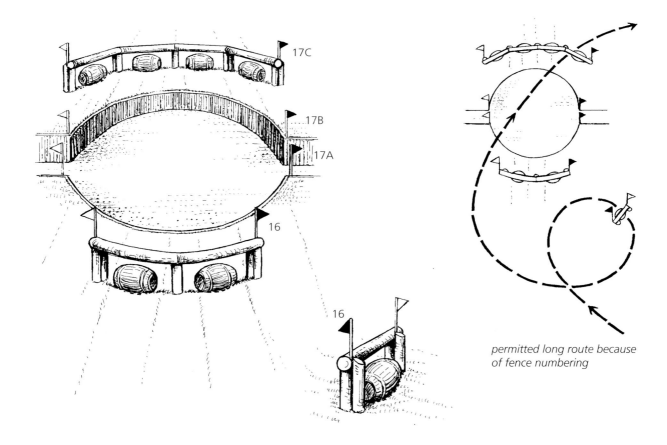

permitted long route because of fence numbering

It is advisable to always have an alternative route for those who may have a refusal or who are stepping up to face such a question for the first time, particularly at the lower levels. The same principles apply as for coffins, and numbering fences separately can also assist with this whilst at the same time allowing the 'flow' and 'rhythm' to remain. Additionally, remember to allow vehicular access to the 'road' part for building and maintenance purposes and also in case there is a need to get an injured horse out during a competition.

When designing Sunken Roads be conscious that all the distances between the various elements are related. It is important to understand how a horse will make its way through this type of question and what happens to its length of stride between the various elements. To mix the distances having some long and others short is not acceptable because of the physical difficulties that this will create for the horses. There is a definite relationship between all the distances in this type of fence and also the height of the step down and the distance in the 'road' element. The exception to this is if 'a' and 'c' are bounces in which case 'a' is 3.05m (10ft) and 'c' is 2.75m (9ft).

Any fence that has a step up to another fence/element must allow the horses to come up that step on a 'moving' stride just as coming up a slope out of a coffin, and this is just as important in a Sunken Road as in any other fence. If in doubt, or the class caters for a variety of experience and size of horse, one can cover all permutations by setting the last element at a very slight angle thereby allowing the rider to pick his/her line to suit his/her horse.

The height of the step down and the distance between the first element and this step will determine the distance in the road part of the fence. Thinking about how the horse jumps, the bigger the step down the steeper the horse will land and therefore the shorter the first stride. This is even more obvious if the horse is being asked to bounce down. These factors are important in working out the distance required to make the fence jump well.

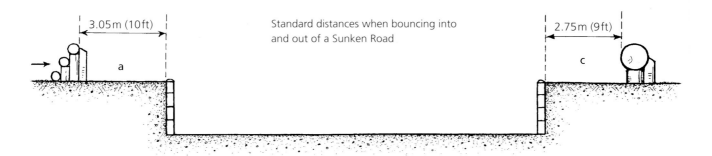

3.05m (10ft)   a

Standard distances when bouncing into and out of a Sunken Road

2.75m (9ft)   c

For instance, a 70cm (2ft 3in) step down with two strides to a step up will require a distance of 9.03m to 9.33m (30ft to 31ft) for a horse whereas if the step down is 1.10m (3ft 7in) the distance could be more like 8.42m to 8.72m (28ft to 29ft) depending on the type of fence and its positioning used for the first element; the difference is brought about because, in the first example, the horse is travelling in a more forward direction when going down the step because it is smaller than in the second example in which there is more downward and less forward movement.

Distances from the first element to the step down will necessarily be shorter than regular distances between elements on flat ground because we want the horses to be closer to the top of the step at take-off. For horses 5.50m to 6.10m (18ft to 20ft) for one stride is usually good, 9.03m to 9.63m (30ft to 32ft) is usually good for two strides; for ponies it will be necessary to shorten these a little more. It all comes back to understanding how horses work and then thinking logically.

To assist the horses it is important to have very clear definition of the steps down and up; the edges must be clearly visible and a small half round on the

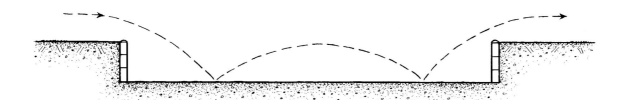

smaller step down means more forward movement in the stride

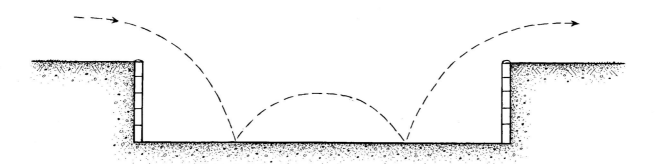

bigger step down means less forward movement in the length of stride

top of the steps will assist with this. Also make sure that there is a clear contrast in colour between the steps, the base of the road, and the ground – we do not want the step to blend into the background with the consequent chance of a horse misreading the question.

As with all combinations, Sunken Roads need to fit into a course. Be careful not to build them where the ground lies wet because of the difficulty in establishing a good base. Also, never have them less than 9m (30ft) length in the 'road' part if there is any chance of there being water in the bottom during a competition. One advantage of building these fences is that there is normally some very useful spoil or fill that can be used elsewhere on the course.

DESIGN AND CONSTRUCTION OF WATER JUMPS

There are normally just one or two options on any given site that are suitable for building a water fence and, with luck, they usually allow a designer a good opportunity to create a fence that will give a variety of options/alternatives. The type of ground and the size of the project will determine the cost and it is possible to spend a huge amount of money unnecessarily building fancy water fences when in many instances a simple circular or kidney shaped 'dew pond' type of fence with sloping sides and jumps around it will suffice for most classes and needs. The exception to this is if the fence has to be specifically created on a site that has no obvious water feature thereby requiring a pond/structure to be built which will hold water.

Dew pond which will allow a variety of options to a designer.

Some useful pointers for water fences are:

- Make sure that there is a reliable water source; if there is not one then be prepared to install one.
- For maximum practical use try to ensure that the jump can be used from all directions and is therefore not tucked up against a fence or boundary.
- Always make sure that the fence can be drained or pumped out for maintenance purposes.
- Make sure that the various routes that horses will take through the fence are well stoned/drained on entry and exit; every time a horse comes out of a water fence it will take approximately 2 gallons (7.5 litres) of water with it and so the ground can deteriorate very quickly if not prepared well.
- There is no such thing as a cheap water fence (unless you are very lucky!).
- Get it right first time; to have to redo, change, or enlarge a water jump is expensive.
- The most important part is the base which has to be right; if this is poor the fence should not be used until rectified (see page 102).
- Have clean water; if a rider falls off he/she should not have to swallow filthy, stagnant water.
- The water depth should ideally be no more than 10cm to 15cm (4in to 6in).
- Any step out of water should be very obvious or not there at all; 60cm (2ft) minimum up to 1.01m to 1.07m (3ft 4in to 3ft 6in) at the highest level – a 1.07m (3ft 6in) step out of water can be a big effort.
- With a jump into water always put 3cm to 5cm (1in to 2in) of sand down where the horses will land in case they peck and go down on their knees; this will help prevent physical damage. Also allow sufficient area to include the first stride.
- When there is a drop into water always slope the landing very slightly down; this will ease any possible jarring.
- When there is a step out of water always slightly slope the approach upwards from about 3m (10ft) before the step to help the horse jump (the same applies on the approach to fences in water); we often see horses misjudge or stumble up steps out of water and whatever we can do to help must be good.
- Always remember that it is very easy to set horses back in their training with bad water fences; riders spend a lot of time getting their horses to

accept this type of fence and we must recognize our responsibility; there is no need to get too clever and we must never produce fences that will cause confusion.

- If possible, allow the horses some time in the water so that the experience is better, i.e. have more than two or three strides if costs/budget permit.
- Before construction begins it is important to know where the sun will be relative to the siting of the fence and when the fence will be jumped; clearly we do not want horses to be jumping directly into a low sun or have any chance of misreading the fence.

There are various dos and don'ts in addition to the above:

- Do ensure that there is absolutely no chance of a horse thinking that it may jump all the water in one go.
- Do remember that a horse's stride will be slightly shorter than normal in water.
- Do not have fences less than two strides apart (9m [30ft]) in water as a minimum, although three strides as a minimum is preferable.
- Do not build a water fence under trees; there are too many shadows that can cause confusion to the horses.
- Do use materials that will last.

See the diagrams on page 100 which illustrate minimum distances.

It is important to remember that the question being asked is 'will the horse jump into water', and that a horse should not be punished inadvertently for doing just that. It is for this reason that there is absolutely no benefit to having water deeper than 10cm to 15cm (4in to 6in) and there is no excuse whatever for a poor base. All bases must be checked regularly depending on the level of usage, certainly at least once a year and before each competition when any sediment build up should be removed and any depressions treated correctly.

As with all fences the profile of those into water is important. Thinking of how horses generally jump these fences it is not rocket science to work out that vertical fences are out of the question and that the shape needs to be forgiving. Horses rarely get too much up in the air jumping into water (except perhaps for green horses in the early stages of training). Generally they are keen to get their feet down and this then tends to have them fairly low over the fence that they are jumping which means that a rounded shape/considerate profile is essential, whatever the level.

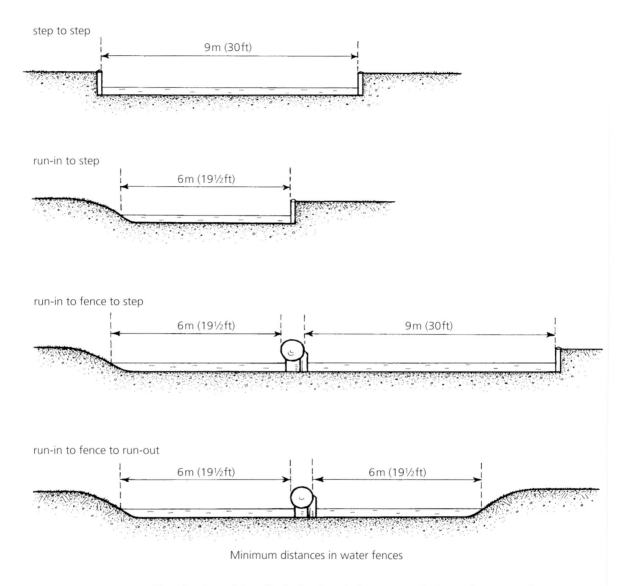

step to step

9m (30ft)

run-in to step

6m (19½ft)

run-in to fence to step

6m (19½ft)   9m (30ft)

run-in to fence to run-out

6m (19½ft)   6m (19½ft)

Minimum distances in water fences

Developing this a little further it becomes obvious that spreads into water are not appropriate questions.

The type of question that is suitable depends on how it fits in to the overall balance of the course. Quite often a better effect, particularly at the lower levels, is to have a fence such as a small log positioned at the top of a gentle slope with the water three or four strides after it in the background, rather than jumping directly into it. We often see green young horses having one refusal and then going the second time which is good education without being frightening or risking losing their confidence. For those who are fortunate to have two water fences on a course it is good to have one fence like this and

then the next one can be a jump into water – this way there is a definite progression in education and confidence giving.

The height of the fences into water is critical. It is very easy to make fences too difficult and too demanding, 5cm (2in) can make a huge difference in severity and if in doubt, it is important to err on the side of being too small than too big. Again, question what is being sought and whether what is being offered is suitable for the competition – this is important.

A word of caution on fences actually in water – be conscious of shadows at all times during the day in case a false groundline appears through a shadow on the water. These are fences that we need to be very careful about. It is not a fence that comes in at the lowest levels but when such a fence is used one suggestion for helping overcome any shadow issue is to put some flowers or similar about 30cm (12in) high where a groundline would normally go. Another hint for fences in water is to have some flowers or brush approximately 10cm (4in) high, along the top of the fence to encourage the horses to get up off the timber; the reason for suggesting flowers is that they will stand out and give

An example of a good shape/ profile fence going into water that is inviting and has good definition.

good definition. It is also very important to ensure that the solid part of such fences are well below (at least 10cm/4in) the maximum permitted for the particular class.

Often it is very useful to be able to use any spoil that comes out of a water fence to build a bank or create steps; this is absolutely fine provided the spoil is a) left to dry thoroughly before being used, b) is compacted regularly as the bank is being constructed, and c) is suitable for this purpose, for instance, does not become deep when wet. Further information on banks is included on pages 27–29.

The principles of constructing the base of water fences are fairly simple and obvious; the important point is to not take short cuts and, if in doubt, ask an experienced course builder rather than guess.

The base of the fence must be secure, this is not up for debate, and the construction process depends on what there is to start with. It is ideal if the ground is such that there is a hard base already; it is then just a case of settling on the depth of water and adjusting the base, the control mechanism (if any), and the water inlet accordingly. Do ensure that the base is suitable by putting horses through it to test thoroughly that it does not deteriorate or break down; never assume that a base will be all right because it looks good. Similarly, if there is a need to increase the area of the fence, do not assume that where the ground will be excavated will also be suitable – it may be necessary to dig out some ground and prepare it correctly. Another problem that has cropped up at several events where a stream is used for the water fence is that the base has deteriorated over a period of time. What started off as a solid base has, through wear and tear, collapsed and the clay beneath the original top surface has

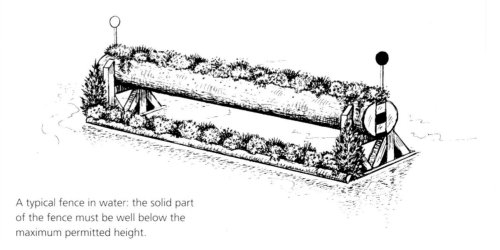

A typical fence in water: the solid part of the fence must be well below the maximum permitted height.

come through. This highlights the need for constant maintenance and thorough monitoring.

Starting from scratch or developing a pond or lake is a different exercise altogether. In order to get a good base all soft material such as mud, sludge, or similar, must be removed and replaced with hardcore. It will be necessary to empty/pump all the water out (and, very importantly, keep it out until the construction is completed) or divert the water course using pipes or temporary drainage lines. Often it is best to work backwards in the planning: work out where the finished water level will be, agree the depth of water that is wanted, and this will then give you the top of the base. Then it is a question of working downwards adding the various depths of required materials to give a starting point from which to start preparing the base. Before starting it is necessary to do a test dig with a JCB/backhoe or similar to establish what the ground is like where the jump is intended to be: it may save a huge amount of money or steer the designer to another location.

If when this point of excavation is reached the material is still very soft it will be necessary to keep excavating until firmer ground is found or for at least an additional 65cm to 75cm (26in to 30in).

At this point it will be necessary to put a membrane down such as terram. This is a threaded porous membrane which allows water to go through but prevents any mud coming up to mix with the stone which is something that is definitely not wanted. Ensure that the membrane is overlapped at the edges when rolled out. Then put 25cm to 30cm (10in to 12in) of 'clean' hardcore or rubble or 15cm to 18cm (6in to 7in) stone – 'clean' means no dust, earth, or mud – onto the membrane. Then add a layer of 10cm to 15cm (4in to 6in) of clean stone followed by 5cm to 8cm (2in to 3in) of 20mm down to dust. At this point the 'dust' is essential for its binding qualities.

As the top two layers are put on it is essential to compact them thoroughly to ensure that they bind before the final covering of 3cm to 5cm (1in to 2in) of coarse sand/limestone dust is put on. This final layer will act as a cushion should a horse fall or trip and go onto its knees and will hopefully reduce the risk of injury. It will need checking, raking, and probably topping up on a regular basis, and do remember to build in a properly constructed vehicle access for maintenance purposes as well as a possible need to get to an injured horse. The diagrams on page 104 illustrate how to set about constructing a suitable base.

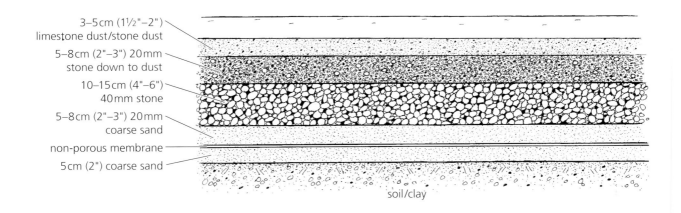

3–5cm (1½"–2")
limestone dust/stone dust

5–8cm (2"–3") 20mm
stone down to dust

10–15cm (4"–6")
40mm stone

5–8cm (2"–3") 20mm
coarse sand

non-porous membrane

5cm (2") coarse sand

soil/clay

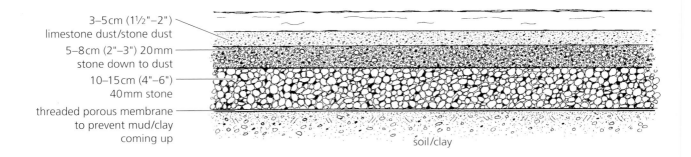

3–5cm (1½"–2")
limestone dust/stone dust

5–8cm (2"–3") 20mm
stone down to dust

10–15cm (4"–6")
40mm stone

threaded porous membrane
to prevent mud/clay
coming up

soil/clay

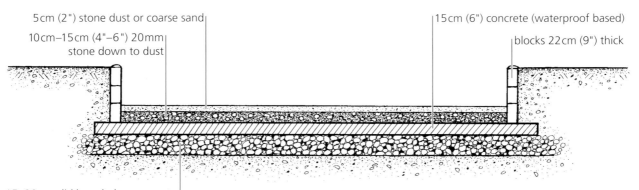

5cm (2") stone dust or coarse sand

10cm–15cm (4"–6") 20mm
stone down to dust

15cm (6") concrete (waterproof based)

blocks 22cm (9") thick

15–20m solid base below concrete

(i) render the blockwork

(ii) seal the walls with swimming pool waterproofer

re (i) there is a waterproofing liquid which can be added to the render

The photographs above show the base of a water fence during construction.

It is possible, but often more expensive, to use 15cm (6in) deep concrete to form a foundation for the base and then do some blockwork around the edges capped with strong hard timber or sleepers. There is the need to then seal the blockwork to retain the water and add the base of 10cm to 15cm (4in to 6in) of 20mm down to dust as previously described with 5cm (2in) of sand/stone dust. The problem with these is that they do not look very natural but at least they work.

Another method for creating a water jump where there is the need to retain water is to use some heavy duty butyl rubber sheeting (good quality heavy duty swimming pool lining can also be used). Dig out the hole to the required depth based on the above criteria. Put down the heavy duty butyl rubber covering the whole area including far enough up the banks so that it ends above the eventual water line; it is necessary to put a 5cm (2in) layer of soft sand beneath it; then put a 10cm (4in) layer of soft sand on top of it before adding another 15cm to 20cm (6in to 8in) of 20mm down to dust which is carefully compacted. This should work well but it will need regular checking to make sure that the butyl rubber remains well protected.

Any revetting of steps needs to take place before the liner is put in since there is no wish to have to make holes in it. To make life easier it is much better to present a flat face of any revetting for securing the liner and so any uprights should be inside the step, out of sight, with the sleepers or whatever is used for the step bolted to the uprights. The liner obviously needs to come up above the proposed water line and it can then be covered by half rounds, also secured to the revetting above the water line.

As mentioned earlier it is very helpful to the horses if the base can be slightly ascending from about 3m (10ft) in front of a step out of water. It is always unfortunate when a horse 'misses' when jumping up a step and anything that can be done to assist the horses to make a clean jump should be done. Likewise it is helpful to grade the landing of a drop into water to help take the sting out of it. All that is needed is a gradual grade up of 5cm to 8cm (2in to 3in) from about 2.5m to 3m (8ft to 10ft) from the step and the opposite when jumping in. Another important issue is to bear in mind that there is a lot of splash in front of the horse and so it may not 'pick' the fence clearly; there must be a contrast between the colour of the water and the step out and if this is a concern it is very easy to paint a white line 10cm to 15cm (4in to 6in)deep around the top of the step to help overcome the problem.

When constructing water fences it is important to test the amount of time

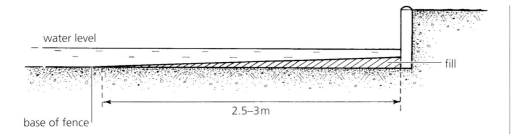

water level

fill

2.5–3 m

base of fence

that they take to fill and that they do in fact remain full for the duration of the competition. Always be prepared to top up when necessary and keep the supply available and primed during a competition.

As with many fences keep the questions for the horses simple and understandable; put the onus on the rider's ability to ride and make them have to consider their approach, presentation, and line.

LOG AND CORDWOOD PILES

These are classic cross-country fences with a variety of looks depending on personal preference and the time available. They can be portable or permanent and they can look good or they can look dreadful.

Whichever is preferred, attention needs to be paid to the profile so that the back of the fence is slightly higher than the front, as per an oxer, and also to how it is built. Clearly there must be no foot traps, all the materials must fit snugly together, and the horses must be able to see the back of the fence as they approach.

For whatever reason log and cordwood piles do not jump as well when the timber is laid end-on to the direction of approach without there being a top rail – horses are inclined to clip them. For this reason it is better to build them as per the diagram by creating a frame of the intended dimensions and then filling it in snugly as required ensuring that there are no foot traps. It is similar to building a parallel and then filling it in. The top rails of this frame should be the usual size rails e.g. approximately 20cm (8in) diameter and there will probably be a need for a groundline of sorts to improve the profile of the fence – this will need assessing when the fence is built.

Framework for log pile before filling.

Building a 'rollover' log pile is straightforward. Create a frame (three or four matching sections depending on the width of the fence) using treated 10cm x 10cm (4in x 4in) timber in the shape that is wanted and then clad it horizontally with either half or full rounds ensuring that they all sit close to each other (it may be necessary to straighten the edges with a saw). To finish the fence off neatly it looks good to put some offcuts of ends of rails at each end to hide the gap underneath it.

If the plan is to make the fence portable then simply build the frame in the desired shape and dimensions (allowing for the rails to be added). When building this as a portable remember to allow for the tractor forks to be able to get under it at the appropriate places for ease of transportation/movement. Portable log piles are extremely useful since they can be changed or moved to create very differing questions – for example, they can be straightforward, they can be related, they can be used to ask turning questions.

Rollover log pile seen end on

Portable log pile frame

109

TRAKEHNERS

It is a good idea to introduce these at the lower levels so that horses (and riders) get used to them and they do not therefore become an issue when suddenly they are seen on a course.

A simple introduction is to create a shallow scoop without much spread under a rail rather than worry about creating a fully revetted ditch and if, even then, there is a concern that the scoop may be too awe-inspiring it is always possible to hide it a little by putting a bigger take-off rail at the front than originally planned. The desire is to get the horses to jump the fence without being intimidated.

Scoop ditch with take-off rail which saves revetting the ditch yet gives the same impression.

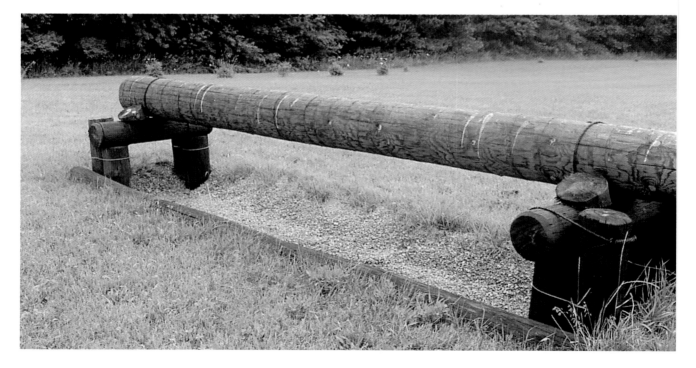

As the levels progress so the ditch will grow from a scoop in depth and spread and have more influence on the look of the fence as it becomes increasingly obvious. There is no real need to be deeper than 90cm to 1m (3ft to 3ft 3in) even at the higher levels.

There is always a discussion about the angle of the top rail and in order to give maximum flexibility it is advisable to build an 'H' frame which allows the angle to change and the positioning of the rail can also adjust backwards and forwards. The important point is that the fence will essentially be an ascending one with the top rail positioned in such a way to assist the horse and feed it in the right direction. For example, if the course goes to the right after the fence then the angle should be such that it leads the horses that way i.e. the right hand end of the top rail is nearer to the take off than the left hand end.

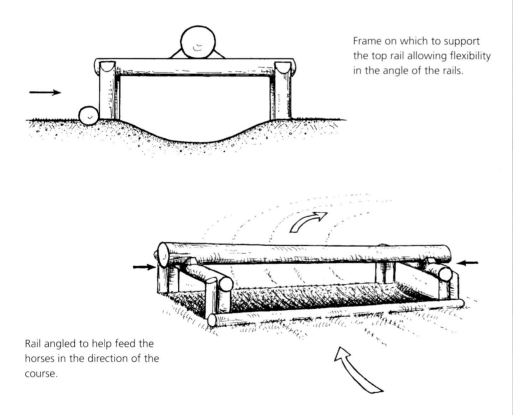

Frame on which to support the top rail allowing flexibility in the angle of the rails.

Rail angled to help feed the horses in the direction of the course.

The diameter of the top rail will depend on the size of the fence and the level of competition but it must not be too skinny. It needs to be inviting and will necessarily be larger than regular rails and anything less than 25cm (10in) will be too small to carry the amount of daylight that there is in these fences.

If it is not possible to obtain larger materials then it is always possible to put three rails together to form a top rail. The bigger the fence the bigger the diameter of the rail should be.

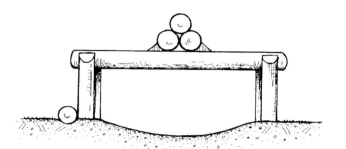

How to use some small rails to obtain the necessary diameter of the top rail.

If it is necessary to revet the landing side of the fence in order to support the ground it is important to revet to 15cm to 20cm (6in to 8in) below the ground level and then round up the ground on top of the revetting – this is so that there is no way that a horse that lands short can injure itself.

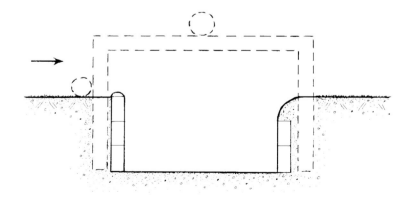

When building trakehners always bear in mind that a horse may get into the ditch and that there must be an easy way to extract it should this happen. Ideally the ends of the ditch are not revetted or if they have to be then it is important to have a ramp at each end to walk a horse out whichever way he is facing. Also recognize that the horse will need to either get out from under the rail, or come up in front of or behind it.

Along the same lines deep ditches, that is more than 1m to 1.10m (3ft 3in to 3ft 6in) are not necessary and should be avoided, and narrow ditches are unacceptable because of the impracticality of dealing with a horse that may get into it; it all comes down to minimizing the risk to horse and rider.

As with all ditches ensure that they are definite and will be understood by the horses. There needs to be some contrast so that the ditch will stand out from its immediate surroundings and this applies to all ditches. An obvious example of what is not wanted is a grassed-in ditch with a grass take off and landing.

ZIG ZAGS

This is another example of a fence that should be introduced at the lower levels whether on its own or over a slight scoop. Bear in mind that some competitors will jump across them, i.e. treat them like a trakehner, whilst others, generally the more experienced, will jump them straight. If there are parts of the fence that you do not wish to be jumped for any reason simply block them off using trees.

At the lower levels the need for much spread is somewhat irrelevant; the aim is to introduce horses and riders to the fence rather than get too carried away, but as the levels progress so can the spread. Zig zags jump better with a scoop or ditch underneath them and it is not good to revet the landing side for fear of a horse not jumping out over the fence, landing short, and possibly

Zig zag over a scoop with good take-off rail and not much spread.

injuring itself. It is also preferable to have a take-off rail on all zig zags, even those not over a scoop, to avoid the possibility of a horse getting too deep; clearly those over a scoop must have a take-off rail.

The width of the jumpable sections is important. At all levels 3.05m (10ft) is a good guide; this can be slightly less provided the proportions of the fence remain good and the horses are not 'squeezed' or 'sucked into' the narrow part of the fence, and it can obviously be increased also. The desired spread will generally determine the jumpable width but if in doubt lay the fence out before any construction begins – it must be fair and may well need three, or even four, jumpable sections to look and feel right.

As with oxers check the profile of the fence carefully; usually it is necessary to have the ends of the rails at the back of the fence a shade higher than the ends at the front of it so that the horse understands the shape of the fence.

The construction of zig zags is just like that of corners and a useful tip is to place a small tree at each upright when they are used in competitions.

Diagram showing where to position the uptights

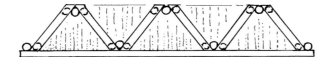

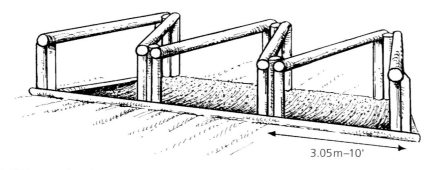

3.05m (10ft) is a good guide as a minimum width of each part of such a fence

CORNERS

These are very good fences to see how well trained, confident, and honest horses are and to test the rider's ability to present his/her horse correctly and accurately. They also provide an ideal opportunity for the better horses and riders to benefit over the less proficient by saving time jumping them as opposed to taking a long route which may feature as an alternative.

Another good thing about corners is that there is absolutely no need to try to get too clever; they work as they are. They can be used on their own or they can be part of a combination, and they can, and should, be used at all levels so that horses and riders get used to them. It is important that they are kept in proportion (not too flat which can happen very easily at the lower levels) and are inviting to look at, and the profile must be such that horses are not sucked into the bottom of the fence. This can be done by using a traditional lower rail(s), an apron, some filling under the top rail with a ground line at the base, palisading, etc.

The debate as to whether to make the top solid (using timbers strong enough to cope with a horse that may touch down on the top and then having some dust on top) or not is one that should be had with every corner. When having this discussion if the feeling is that there is a possibility that a horse may end up in it, then fill the top in. If the approach and profile are such that it is very easy for the horse to understand what it is being asked then it is preferable not to fill the top. There is absolutely no doubt that most riders will say that they do not respect corners with a solid top nearly as much as an open one and more will be tempted to have a go. For this reason I try not to have to fill them in. If the decision is to platform the top it is a good idea to put some dust or similar on the top that will just give a little bit of purchase for any horse that may bank the fence; do not use anything slippery such as shavings.

All corners must have some sort of construction or barrier to prevent horses inadvertently drifting to where the corner is too wide to be realistically jumped. This can be done by using tops or brush bundles on fences without a platformed top, or for those with the top filled in it is very easy to construct something attractive using offcuts or another option is to put a half barrel or similar on the top.

Whether to put a tree/top on the apex to help the horses jump the fence is

above A good corner with good profile and definition.

left Dust on the top of a corner.

The middle of the corner has been blocked off to prevent horses jumping it where it is too wide.

very much the designer's choice but it seems a shame in some ways to take away the benefit from those that are good enough to jump it without a tree there.

With regard to the construction, always have the back a little higher than the front and always make the profile inviting. Additionally, for 'open' corners the rails should be outside the upright on the apex so that if perchance a horse is in the middle of the fence the back rail can be removed by taking it away from the horse.

From a construction standpoint it is also good to place the upright inside the apex for those corners that will have a solid top. Another advantage is that the ends of the rails can be cut off a bit at a time once the fence is erected to determine the level of difficulty.

When setting up a corner ensure that the line that is available to the horses is correct. Horses must be able to jump a corner on a line that is straight on to a line that bisects the angle of the corner; it should not be asked to jump across this line.

This type of fence is part of the sport and therefore needs to be part of a horse's (and rider's) education, hence the need to introduce them at the lowest

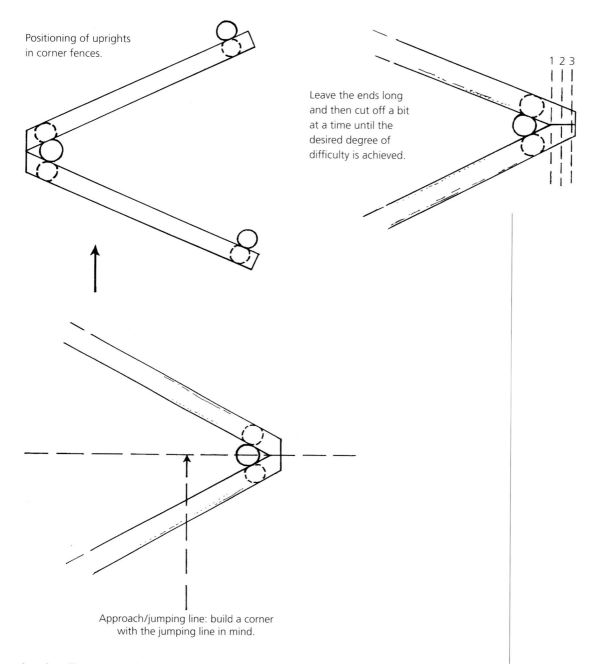

Positioning of uprights in corner fences.

Leave the ends long and then cut off a bit at a time until the desired degree of difficulty is achieved.

1 2 3

Approach/jumping line: build a corner with the jumping line in mind.

levels, albeit in such a simple form that they are sometimes almost oxers. At all levels there should be an alternative route or an option but it is perfectly acceptable, and indeed recommended, that simpler corners are built on a long route so that the question is not avoided altogether.

The method of laying out corners on site is down to personal preference; the line has to be right and the angle needs to be suitable for the level of competition. A guide is 45 degrees for BE Novice (USA Preliminary), 60 degrees for Intermediate, and slightly more, towards 75 degrees for Advanced (we have all seen and designed fences at 4 Star level that are 90 degrees – so much depends on the siting of the fence and any relationship to another element/fence). I have always found the best way is to lay rails out on the ground and play with them until I am happy with the angle and the line; this avoids building the fence and then having to redo it!

Double corners, whether the same way or inverted, are now commonplace on courses from Intermediate upwards. They are not really appropriate at BE Novice level. A useful tip when making a corner as a second part of a combination is to remember that it is sensible to make it a little less wide than the first corner in case a horse is struggling to reach the back of it. Also, ensure that the line being offered to the riders is fair and achievable, not too skinny and not such that they are being asked to jump the corners where they are too wide.

What can help those at the lower levels introduce their horses to a corner is if a simple fence is located four or five strides before it so that the horses are put in exactly the right place to jump it. This reduces significantly the chance of the rider 'missing' and thereby creating a possible problem. It all comes back to the education and producing of horses.

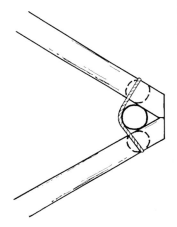

above How to rope a corner.

Double corners.

120

WALLS

It is always fun to be able to incorporate some natural features and walls are common in many parts of the world. They are not necessarily the cheapest fences to produce because they will probably need foundations (depending on size), they will certainly need reinforcing, and they will have to be made suitable to be jumped by building a wooden frame or setting wooden top edges to minimize the chance of a horse injuring itself. The wooden capping on a wall is essential and it must be secured and then cemented in place. It does not matter whether a quarter round or a full round is used provided that it is secure and prevents a horse injuring itself on the front or back of the fence.

If converting a natural wall, rather than just building a frame around what already exists, it will be necessary to recess the capping into the top so that there is nothing 'false' about the fence. If starting from scratch it is a good idea to set up some stringing to get the desired shape and dimensions of the finished product before work starts; then the frame can be built followed by the walling itself. Remember that all walls, particularly those that are just capped

Wooden capping to prevent hooves hitting the stonework.

121

as opposed to 'framed', will need to have some foundations – if in doubt consult a builder/waller but remember to tell him what it is to be used for!

An essential consideration when constructing walls is how to build them so that if a horse should hit it really hard it will not collapse or move, and so any frame must be stout and secure – walls that collapse when hit are not only an embarrassment, they can also cause serious injury.

To finish the top on walls it looks good to set some of the stone used in cement and, in some cases, rounding off the top improves the profile.

right and below Stones set into the concrete on top of a fence to give a good finish. Note that there are no rough or sharp edges.

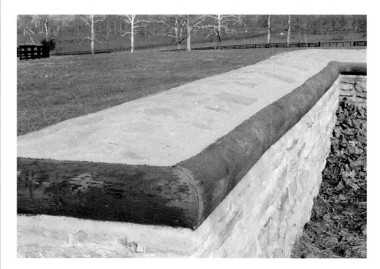

STEEPLECHASE FENCES

Around the world there are many different types of steeplechase fence that are used in three-day events; some are firm, some are very soft, some are timber, some are plastic, it all depends on what is used in the particular country.

In the UK we have a standard brush steeplechase fence that stands at 1.37m to 1.4m (4ft 6in to 4ft 7in) and they vary between the fixed ones and the portable ones which are built in sections 1.8m to 2.4m (6ft to 8ft) wide usually using three sections to make up one fence.

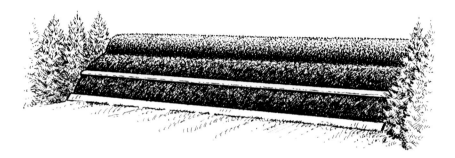

Portable fences obviously give more variation in location and, if they are only used once a year, can be removed from the site for storage and also to free up the land for its usual usage. See the diagrams on page 124 illustrating how to build the fixed and the portable fences. There are several important points that apply to each:

- The angle of the face of the fence should be between 48 degrees and 50 degrees.
- The brush must not be too firm; horses must be able to brush through the top of it, it must not be like a solid fence.
- The timber frame must not be higher than 70cm (28in) and must be bolted together, not nailed.
- Always have wings when used in competition, even if just trees.
- Always make sure that the rails are of a bright colour e.g. white or orange.
- All rails must be secure; for the permanent fences the rail on the belly of the fence can be wired through the apron to the timber frame.
- Always have a ground rail.

- Always have spare material.
- For preference try to avoid siting them on downhill slopes.
- Do not use thick brush; anything more than 15mm (½in) thick should be removed.
- Use treated timber for the frame.

Additionally, for portables:

- Ensure that they are well pegged down.
- Where the sections join make sure that they are flush and secured together.
- Use some good quality small gauge chicken wire underneath the fence to stop the brush falling out when the sections are moved.

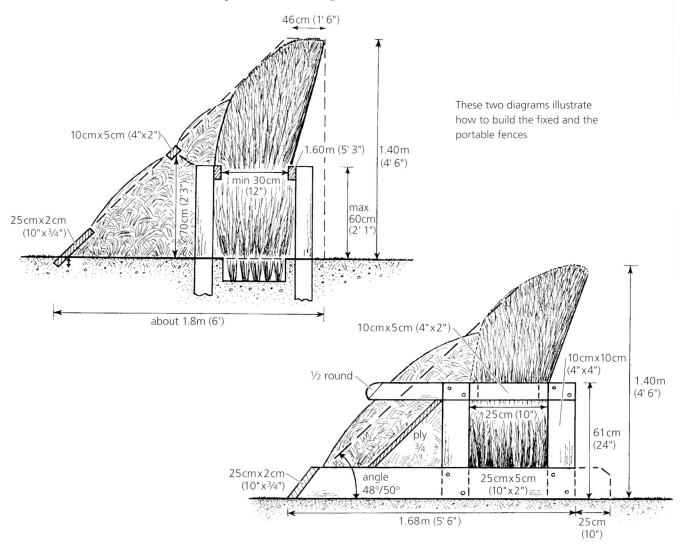

These two diagrams illustrate how to build the fixed and the portable fences

For the permanent fences the diagram illustrates that a trench measuring approximately 30cm (12in) deep and 30cm to 35cm (12in to 14in) wide should be dug for the length of the fence before the timber frame is constructed. The width of this trench must be no more than the width at the top of the timber frame in order to help with the packing of the brush.

When trimming the brush use a good quality brush cutter or hedge trimmer and it is always a good idea to rig up a string line to assist with getting a good top and a good rounded shape. Remember to remove any stemmy bits that could injure a horse. These will appear after cutting. Also check that there are no sharp sticks in the fence.

The firmness of the fence will be determined by how much birch is packed into the fence. When doing this packing it is best to put in four to six bundles (depending on their size) and pull tight using a rope hitched to the back of a vehicle (manpower alone will not be sufficient). Secure it and then add another four to six bundles, and so on until completed.

The brush should be packed before it is cut (it will then look like a big bullfinch) but if it is really tall it may be better to cut off some of the thicker bottom part of the brush so that when it is cut it will not be too thick at the finished height. Check the brush carefully before you decide the best course of action.

Brush does become brittle after a couple of years and will need replacing although much depends on the weather.

For the apron there are several options and they rather depend on the materials available. For permanent fences it is possible to use the offcuts from the brush packed tightly and then topped with gorse or more brush laid vertically before trimming to suit. For the portable sections it is possible to save on the 'padding' by putting some 1.9cm (¾in) marine ply as shown in the diagram on which to create the apron.

For the steeplechase phase of a three-day event the fences should be 6m (20ft) wide so that if the ground deteriorates during the day there is plenty of fresh ground available if riders choose to use it. It is also important that the fences are up to height because they will jump much better. If they are only, for example, 1.10m (3ft 7in) the horses will start hurdling them rather than jumping them which is not good preparation for the cross-country part of the competition.

PORTABLE FENCES

We are all seeing and using more portables nowadays. The advantages are obvious in that there is greater flexibility when changing a course from one year to another, they can be relocated in the event of bad weather, they can be taken in and stored thereby hopefully prolonging their life as well as freeing up the land for agricultural purposes if necessary, and they can also be hired or loaned to other organizers. Initially they will cost more to construct but the belief is that this is more than outweighed by the benefits.

There are some very good international courses around the world that consist almost entirely of portable fences because there is very little lead time for the build up and permanent fences are not permitted. This inevitably means that there needs to be much imagination in the design and construction.

Portables by their very nature are heavy if built correctly. They need to be strong and robust to cope with the equestrian requirements and they must be able to withstand being moved around by heavy machinery.

There is absolutely nothing wrong with portable fences provided:

- They look like a regular cross-country obstacle.
- They are properly secured to the ground; do not be tempted to think that because they are heavy they do not need properly securing; if a horse hits

right and opposite page
Examples of portable fences that are easy to build.

it, it must not move or, even worse, roll over. Officials must always remember to check that all portable fences are secured down.

The frame for portable fences can either be wood or it is perfectly acceptable to use box metal if preferred. The important point for whatever material is used is that the principles of safe construction are adhered to and there are no sharp edges.

·

PART FIVE

Putting Design into Practice

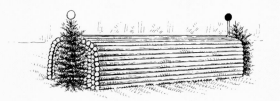

'Are we achieving what we want to achieve?'

ANALYZING/EVALUATING COURSES

We have looked at the theory of course design and the next step is to put it into practice and start to get a feel for what is being looked for. This is an exercise carried out by many designers to get an overview of a course and the number and type of fences planned or used, that is to say the number of uprights, drops, narrow fences, etc, to ensure that no particular type of question is overdone and that there is enough variety on a course.

Taking this further, let us use some examples of existing courses to consider their 'balance' and 'feel'. It is difficult without knowing the terrain but the principles are still there to be debated. Assuming that the finished heights of the fences are suitable for the level of competition the issues to consider are:

1. Are there enough questions and is the course suitable for the level of competition?

2. Is there a good variety of fences and questions?

3. Would the horses and riders benefit from riding the course?

4. Are we achieving what we want to achieve?

5. Are any particular types of fences over-used?

6. Are we asking questions to the right and to the left, not all in the same direction?

COURSE A. NOVICE ONE-DAY

1. *Log* Very inviting straightforward fence 5cm (2in) below maximum to give confidence to these less experienced horses. The good shape makes it easy to jump and the ground is very slightly rising on the approach.

2. *Chair* Another straightforward fence this time at maximum height offering a good shape and good approach continuing the thinking behind the first fence which will also give the horses and riders the opportunity to settle into a rhythm. We want the horses to be enjoying the experience.

3. *Sloping Rails* Ascending fence landing on a gentle down slope which is designed to gently introduce the concept of a drop fence. The fact that this comes after a left bend will mean that the horses will not be going too fast should the rider's enthusiasm get the better of him/her.

4. *Parallel* This is the first fence at maximum height and spread; it is very inviting with the three rails on the front to make a good profile and comes

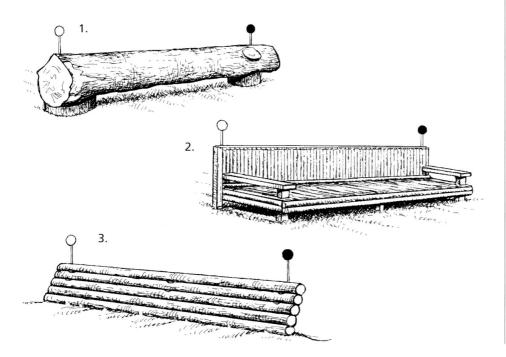

fairly soon after fence 3. By now the horses should be up in the air and all the fences so far have a different look to them so that the horses get to see variety. This is used as the first 'stop' fence.

5. ab *Curved Rails* Now that the horses and riders are well into the course it is time that they were asked to jump some type of combination. These two sets of curved rails on two strides on a gentle down slope (both at 1m/1.02m or 3ft 3in/3ft 4in) are designed to ask a variety of simple questions – can the rider present the horse in the correct place in the correct shape at the correct speed on the correct line? It is not a difficult question but it requires the rider to walk the line carefully when inspecting the course. The profile of the two elements is good and will be forgiving should the rider make a mistake and it is also very clear to the horses what the question is – there should be no confusion.

6. *Trakehner* A classic cross-country fence which horses will and should meet at all levels with plenty of width to help the horses and riders and a very inviting profile. It is only 1.02m (3ft 4in) but it looks bigger and the fact that the ditch is just a scoop rather than a fully blown revetted ditch makes it look less intimidating. What we want to do is encourage horses to jump the fence, not eliminate them, and so the take off rail is very obvious which hides more of the ditch than if it was smaller. We also need to be conscious that this is the

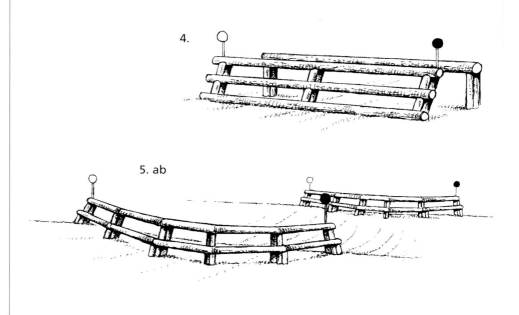

4.

5. ab

first 'ditch' on the course and that at this level there are some horses who may jump 'up' more than 'out'. This means that the base spread should not be too wide lest any horse drops short on landing. Given that this fence is very inviting there is no alternative; a horse at this level should be able to jump it.

7.ab **Corner** Another common fence which has been given a good profile by creating an apron on the front. The apron does not come more than 60cm (24in) in front of the fence at the bottom since we must not be encouraging horses to be taking off too far in front of this type of fence because when they get to the higher levels the back rail will be a long way off. The question is can the rider ride a line, present his/her horse at the correct spot at the correct pace, and is the horse honest enough then to jump the fence? There is deliberately no tree on the apex so that those who are good enough can gain the time advantage over those who are less experienced or less confident of their own or their horse's ability. There is a long route over two elements.

8.ab **Log to Step into Water** The log is set at 98cm (3ft 2in) on a slight down slope three strides before an 80cm (2ft 7in) step into the water which in turn is approximately 18m to 20m (59ft to 66ft) in length and 14m (46ft) in width. Rather than have a traditional fence into water this type of question uses the water in the background so that the horses can see it and decide accordingly

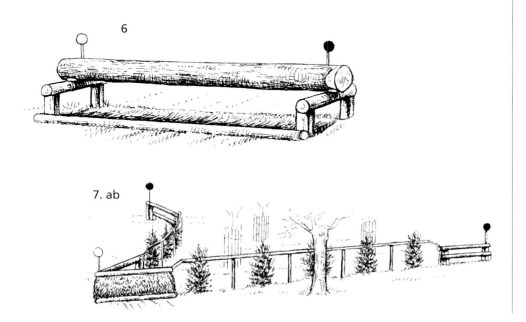

6

7. ab

what they are going to do. It is a confidence question which still gives the greener horses the time and the chance to consider the question in their own time, and the fact that the log is not very big and is also a very forgiving shape, combined with the three strides between the log and the step, means that there is a good opportunity for all horses to benefit and learn. The step down also has a 10cm (4in) half round on the top to give definition. The Fence Judge should always be asked to ensure that this half round is clearly visible at all times and to therefore regularly drag back any stone dust/footing material that may get pushed against it.

9. ab *Step Out of Water to Narrow Rails* The horses have plenty of time in the water which is really good for the less confident horses before they come to the step and rails out where there is a choice of one stride to the rails or a long route. The step is no more than 75cm (2ft 6in) and the water is 10cm to 15cm (4in to 6in) deep; the footing in the water is slightly rising as it approaches the step (refer to Water Jumps page 97), and there is a gentle left turn to the step which will help to ensure the horses are set up. In this instance the rider has a choice as to whether to take the quick or slow route. The quick route is up the step and over the rails which are 1.02m (3ft 4in) in height and 3m (10ft) wide and angled slightly; it is inviting and most horses at this level should be capable of jumping it. The main reason for the alternative is for those who have a

8. ab

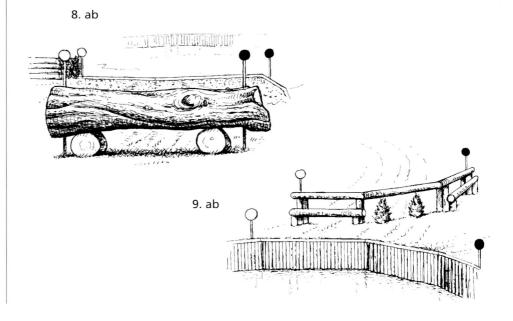

9. ab

run out at the rails because it is not easy to re-present for the second attempt. We must make provision for these situations at all levels.

10. *Zig Zag* This is effectively two sections, each being 3.6m (12ft) wide over a shallow scoop. It is the second 'ditch' on the course, it is up to height but is only 1.2m (4ft) base spread to cater for the less experienced horses who may be suspicious and jump 'up' rather than 'out' as per the thinking behind the dimensions of the trakehner, fence 6. It is a straightforward fence to allow the horses to re-establish any confidence that may have been lost at the water fence. It is also designed to introduce horses to this type of fence if they have not met one before. On this course it is used as one of the 'stop' fences.

11. *Arrowhead* This is positioned towards the end of a long gallop and there is some course roping on each side which prevents the riders sweeping wide and jumping the arms without having to slow down (but it does still permit them to remain in a good rhythm). The idea is to control the speed. The fence has a good profile and the width of the front face is 1.8m (6ft).

12. *Drop log* Slightly light into dark with the next fence in the background the log is not very big at 90cm (3ft). It does have a small drop. The riders need to ride this with the next fence in mind – too fast and the next fence becomes

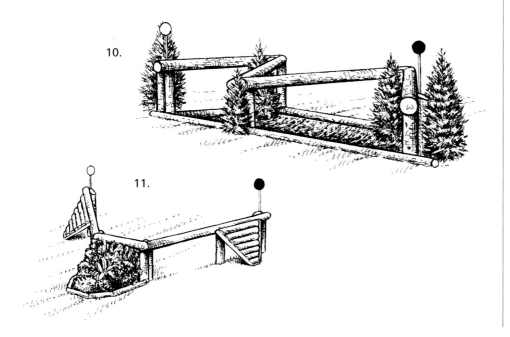

much more difficult because it comes up after seven or eight strides, too slow and a refusal will be a possibility.

13. *Curved Brush Oxer* Very much related to fence 12 the 'scallopping' of the brush makes accurate, positive riding essential. On the quick line the brush is at 1.2m (4ft) and the width of the scallop is 1.6m (5ft 3in); on the long route which wastes time the brush is at 1.15m (3ft 9in) and the width of the scallop is 2m (6ft 6in). Both routes have a top spread of 1.1m (3ft 7in). The quick line is for the confident and the bold and the alternative is there for anyone not so experienced or who may have a run out trying the quick line.

14. *Log Pile* A straightforward fence of maximum dimensions with a good profile and shape that can be used as a 'stop' fence. It is sited approximately 50m (55yds) before the next fence.

15. abc *Angled Rails* One stride between each element in the centre, one and two strides on the left, two and one on the right which gives some choice for the riders depending on their own preference to suit their particular horse. This type of fence is good for this level of competition; it gives the horse

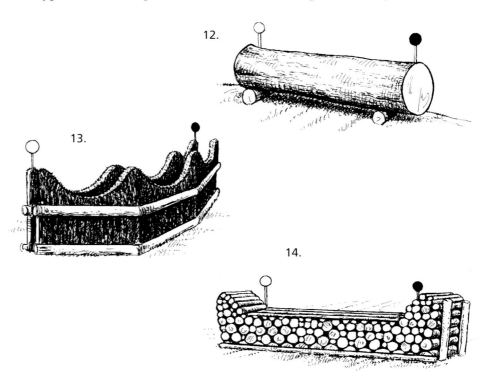

plenty to think about without being unreasonable, it makes the riders think about the distances carefully as they walk the course to decide their best line, and there is nothing that should upset the confidence of the horses.

16. **Oxer** A straightforward fence of maximum dimensions that firstly gives any horse that may have had a run out at the previous fence the chance to settle down again, and secondly, it is the last 'stop' fence on the course.

17. **Ditch and palisade** Slightly uphill, the ditch is 45cm (18in) deep with a 1m (3ft 3in) inside measurement, and the palisading is at 1.1m (3ft 7in) height. It is very inviting and by this stage of the course the horses should be used to small ditches being incorporated into the jumps.

18.ab **Eyelashes** This is the last proper question on the course. It is two sets of curved rails on two strides with an accuracy line on the direct route and a time-consuming long route. This is again slightly uphill, both sets of rails are at 1.1m (3ft 7in) and have a good groundline set 60cm (24in) in front of the rails. The question is set to see if the horses are still paying attention at this stage of the course and to check that the riders are not getting too carried

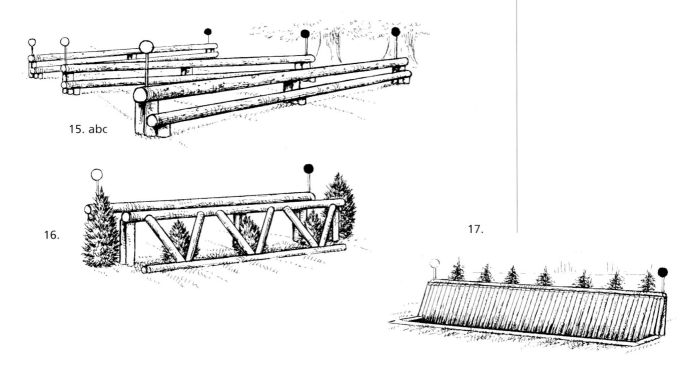

15. abc

16.

17.

away. It is also there to control the speed so that the gallop for home does not start too early. It is the sort of fence that horses and riders would jump time and again at home with no trouble but in competition, for whatever reason, it does cause the odd run out.

19. **Sheep Feeder** The first of two straightforward fences at the end of the course that will hopefully allow everyone to finish full of confidence. It is up to height and has a top spread of 1.18m (3ft 10in). It is set under some trees not too far after fence 18 and so the horses will not be travelling too fast.

20. **Flower Box** At maximum height with a base spread of 1.53m (5ft) off a gentle right hand bend the plan is to finish them well.

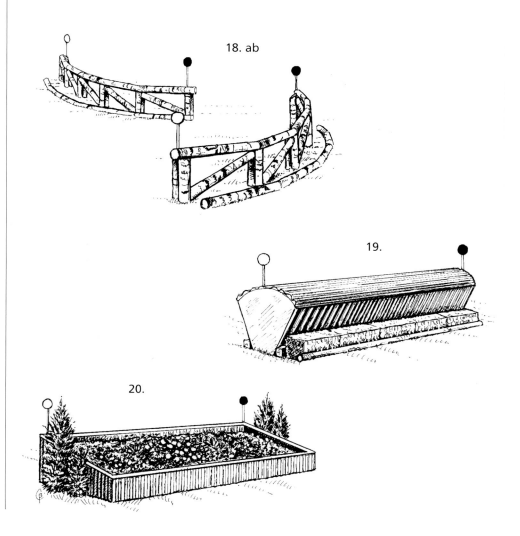

No of Jumping Efforts – 26 (corner counts as one provided this is where we expect the average number of competitors to jump).

ASSESSMENT

Good variety and broad selection of fences, a couple of accuracy questions, several ditches incorporated, two simple drops, both onto ground that is slightly descending, some fences where riders have to make decisions on striding. The design gives a good opportunity for riders to get into the course and for horses to grow in confidence; five combinations/related fences. Speed is taken into account with the design, there has been an attempt to make the fences look different and the horses should learn and benefit from their experience. It is well laid out for the expected management and administrative issues, for example, there is easy and good access for emergency vehicles.

COURSE B. ADVANCED ONE-DAY

1. ***Ascending Spread*** An unusual looking fence which is fine for this level (not something that would be necessarily appropriate at Novice level) at 1.2m (3ft 11in) height and 1.8m (5ft 11in) spread. It is inviting and will get the horses in the air.

2. ***Stick Pile*** Maximum height with a top spread of 1.75m (5ft 9in). Straightforward at this level but getting into the course more quickly than at the lower levels. With this type of fence we always need to evaluate whether we feel a horse could get into it and therefore have a game plan to get it out; the alternative is to fill the top of the fence.

3. ***Log Height*** 1.2m (3ft 11in) slightly uphill set back from a small lip in the ground. With the uphill approach it jumps very well and the brush set at 1.4m

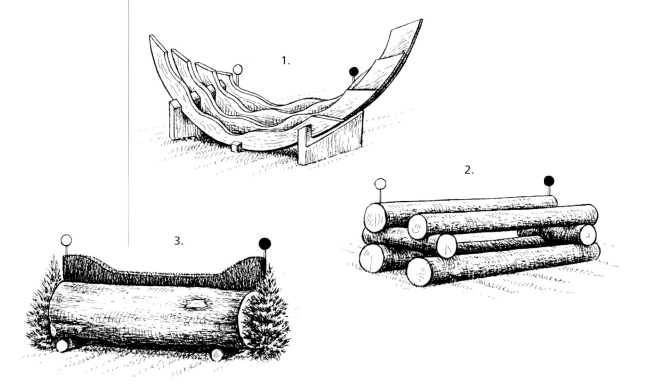

(4ft 7in) also encourages the horses to get well in the air. First 'stop' fence on the course.

4. *Table* Off a left bend this is a big fence. At 1.15m (3ft 9in) it is not maximum height nor at 1.5m (4ft 11in) is it maximum spread; it does not need to be any bigger. It is over a scoop in the ground and the approach is very slightly down hill. The key to this fence is its shape and its front; it must not suck the horses into the bottom of the fence.

5. ab *Two Arrowheads* Having opened up the horses over some big fences it is now time to ask a control question. Both elements are set at 1.2m (3ft 11in), the two elements on the direct route are 1.4m (4ft 7in) wide, and there are three moving strides in the middle. It is also slightly down and then up in between and there is a long route available mainly for those who may run out. This question is one of accuracy and honesty – a good stride at the first element will make things much easier for the horses at the second element because if they are a little way off part 'B' the inclination will be to run out.

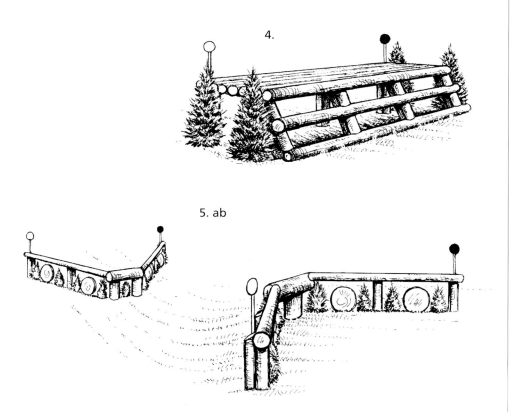

4.

5. ab

6.ab *Corner (left hand)* Big fence with a good profile. There is a long route over two less wide corners for the not so bold.

7. *Hammock* Set on a piece of flat ground as the course works its way up hill this is at maximum height 1.2m (3ft 11in) with a spread of 1.6m (5ft 3in). It is not at maximum spread because of the combination of the facts that a) the course is working its way uphill even though the fence is set on a small flat area, and b) if a horse should be a little too far off the fence at take off we do not want it not making the spread. It also has a solid top in case any horse should put a foot down; an open fence here would not be appropriate.

8. *Oxer* This is set six strides before the next fences. It is maximum height and spread and riders need to jump this with the next fence in mind. It should help set up for the next fence. This is the second 'stop' fence on the course.

9/10. *Bounce* Set at the top of the hill there is a 90 degree turn to this question from fence 8. The two elements are set at about 1.05m (3ft 5in) and there is a slight drop after the second element. Fences 8, 9 and 10 are a good test of

6. ab

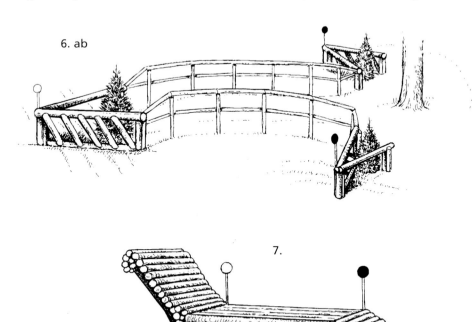

7.

riding ability and horses' athleticism and confidence. The profile of the two elements of the bounce are encouraging the horses to jump it well. There is a long route for the less experienced or committed and it is numbered separately so that the long route can flow well.

11.abc *Coffin* Log, one stride to ditch, one stride to log. 'A' is at 1.15m (3ft 9in), the ditch is 1.38m (4ft 6in) inside measurement, and 'C' is set at 1.20m (3ft 11in). There is no alternative. From the previous fence there is a down hill run before swinging left handed to this fence. The important issue here is that the horses can understand the question, and so the height of the first element is such that the horses can see the ditch over the top of it in good time and are able therefore to make a decision as to whether to jump it or not. The real

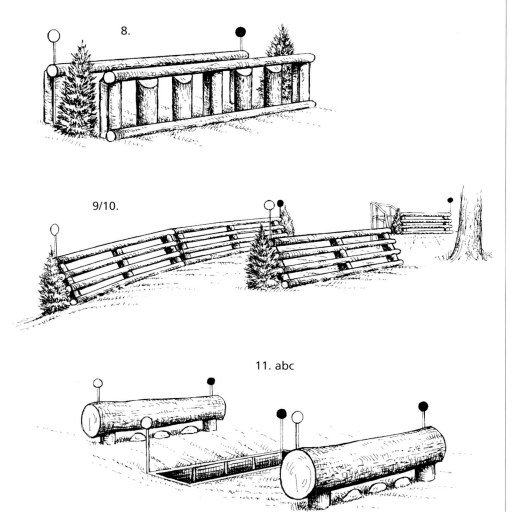

question is at the first element; once over this the rest should be straightforward for horses at this level.

12. *Log* This is a large diameter log set at 1.1m (3ft 7in) coming down a slight slope with the water in the background. The back of the log is approximately 3.6m (12ft) from the water's edge and there is a slight drop. At this point there is a lot for the horses to take in and assess. The pace of the approach will be all important in that the riders will need to give the horse time to assess the question. It looks as though the log is straight into the water until the last two strides. Additionally there are a lot of spectators in the background. In situations such as this 5cm (2in) can make a big difference in the severity of the fence and it is important to recognize this. There is no alternative, the feeling being that horses at this level should jump this fence.

13. *Log in Water* Having jumped into the water the course then bears left and the water stretches out for approximately 50m (55yds). This fence is set in the middle of the water. It is a log with 20cm (8in) flowers along the top to encourage the horses to get well up over the fence and a good ground line of flowers to give good definition through the spray that will be created by the horses. The log is set at 1.1m (3ft 7in) and allows at least twelve strides for the riders to get organized having jumped the previous fence.

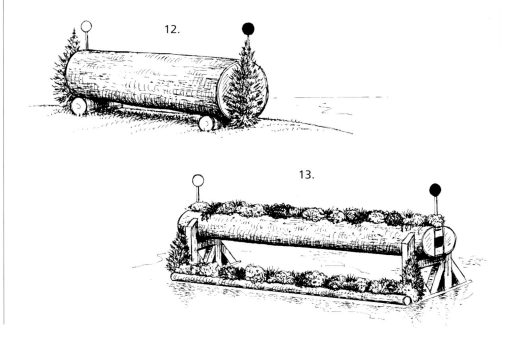

14. **Seat** A very simple fence at 1.2m (3ft 11in) set on a small platform going up hill after leaving the water. A let up fence for anyone who may need a confidence boost.

15. **Brush Oxer** A big brush parallel at maximum height and spread. It is a straightforward power fence that comes shortly after a brief climb and can be used as a backup 'stop' fence if needed.

16.ab **Upright to Skinny** A control and accuracy question in some trees; the first element is at the top of a slope and the second element is set eight or nine strides after 'A' having come down quite a steep slope. There is an alternative to 'B' for anyone who may run out or who may not be too confident at narrow fences. This is a good opportunity for the good horses and riders to gain an advantage over the less experienced. 'B' is set on flat ground two to three strides from the bottom of the slope and it has a small scoop ditch in front to prevent anyone getting too deep should they be very much on their forehand coming down the slope to it.

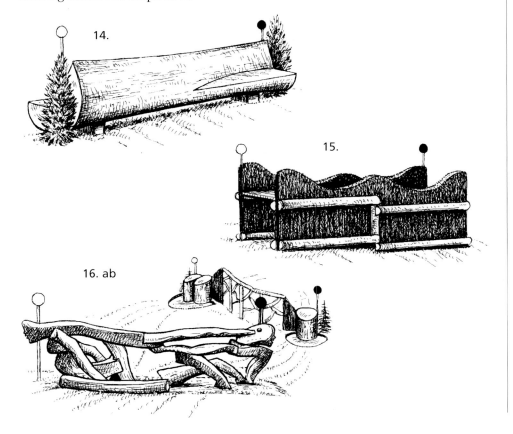

17. **Dray** Another fence at maximum height and spread that acts as a 'stop' fence before the next water.

18. **Log into Water** This asks a different question to the previous water fence; this time competitors are asked to jump straight into the water over a 90cm (3ft) log with a drop of 1.7m (5ft 7in). Confidence and athleticism combined with quick thinking are needed since the next fence comes up after five or six strides. There is no alternative. If the horses were a little hesitant at the previous water fences they may look at this but in reality the length of time that they had in the water at the earlier fences should help them here.

19. **Sluice Gate** Set in the water the way this jumps will be dependent on how well the horses have jumped in. It is 1.1m (3ft 7in).

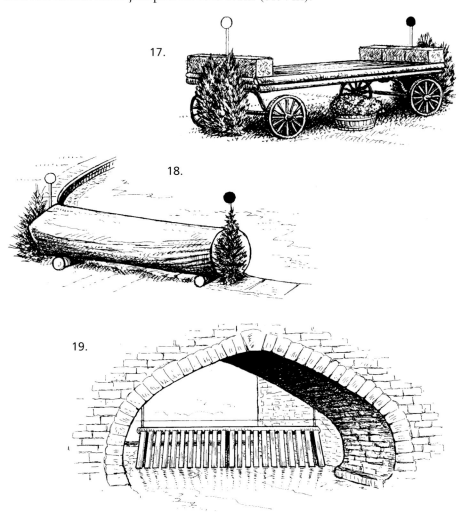

20. ab ***Step up to Log*** Another athletic and power fence up the step and then bounce over the suspended log at 1.15m (3ft 9in). There is an alternative to 'B' for anyone who has a refusal. This alternative means that competitors do not have to go back into the water for a second attempt.

21. ***Picnic Table*** At maximum height and top spread this is a confidence booster after the water questions but still insisting that the horses keep jumping.

22. ab ***Double of Log Piles*** Slightly angled so that riders can choose their best line to suit their own horse(s) both are at 1.2m (3ft 11in) with top spreads of 1.7m (5ft 7in) and set on two strides. Not a difficult question for horses at this level but needs respecting.

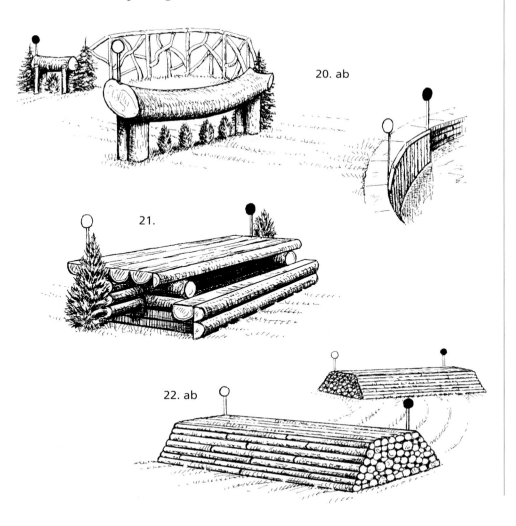

20. ab

21.

22. ab

23. ab *Rails to Flower Box* The rails are sited at the top of a dip in the ground and are 1.15m (3ft 9in) high with a good groundline. There is a slight drop on the landing to sloping ground and then the Flower Box is at the top of the slope coming out of the dip; it is 1.2m (3ft 11in) high, 1.6m (5ft 3in) wide at the back, and has a base spread of 1.6m (5ft 3in). There are approximately five strides between the two elements and there is an alternative to 'B' which is not so narrow and therefore offers less chance of a glance off. The question is all about correct presentation at the first element which is plenty big enough and then we are looking for the competitors to jump a narrow fence on a moving stride. It would be very easy to have a run out at 'B'.

24. *Sloping Rails* An inviting drop fence onto descending ground. This fence needs respecting. The shape is all important and when setting this type of fence it is best to have it high to start with and then take it down a bit at a time until the desired height is reached. 5cm (2in) can make a big difference – too high and it becomes unfair, too low and it is too easy. This fence is at just over 1.1m (3ft 7in) and it looks big enough.

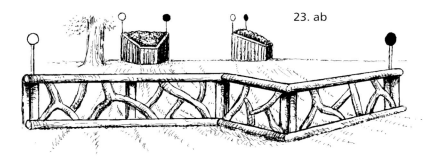

23. ab

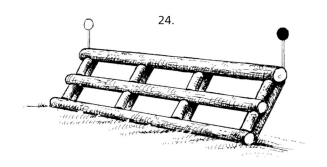

24.

25. **Wall** A simple fence shortly after fence 24 it is set at 1.2m (3ft 11in). At the end of the course starting to unwind the horses and riders but still keeping them jumping.

26. abc **Ditch and Palisade to Corner** The intention behind this fence is to make the riders pay attention and keep riding. The brush on the first element is at 1.4m (4ft 7in) (palissading at 1.15m [3ft 9in]) and the ditch is 1.2m (3ft 11in) inside measurement, 45cm (18in) deep, and then there are four strides on a gently curving left handed line to the corner which this time is right handed and 70 degrees. It is also platformed in case anyone makes an error on their riding and has their horse too far off it. The advantage of having a curving line is that riders can make their own distance; they must also walk this fence carefully to ensure the correct line and approach to the first element. A run out at 'B' would be very easy. There is an alternative to the corner.

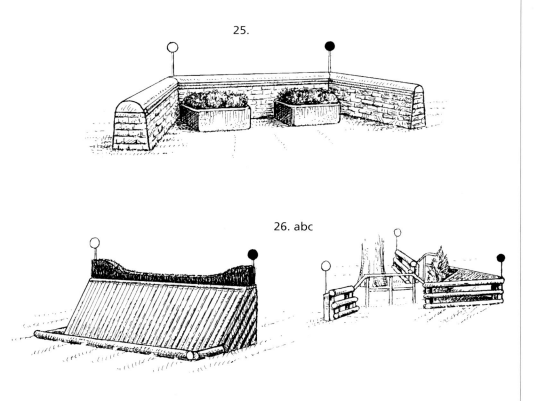

25.

26. abc

27. *Oxer* The last fence is at maximum height at the back and 1.15m (3ft 9in) at the front; it is not maximum top spread at 1.7m (5ft 7in) but is imposing enough to command respect and keep horses and riders switched on. The ground rail is pulled out 60cm (24in) in front of the fence.

27.

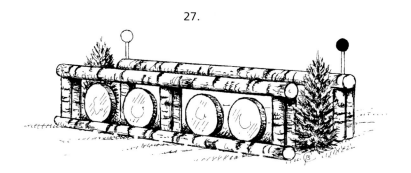

No. of Jumping Efforts – 35

ASSESSMENT

Good variety of types of fence, up to height, more sophisticated questions than course A; there are several fences where riders have to plan their routes and walk the course carefully. Good rider questions are asked and accuracy questions not overdone. The designer has been careful not to have too many drop fences, there is a good opportunity to get going on the course and it keeps riders working to the end. There are eight related fences/combinations. Horses and riders still have the opportunity to get into the course and should benefit from going around it.

COURSE C. CCI***

1. *Flower Bed* Inviting first fence up to height at 1.2m (3ft 11in) and with a 1.8m (5ft 11in) base spread it is designed to get the horses jumping and paying attention right from the start.

2. *Log Pile* Another straightforward fence at 1.2m (3ft 11in) this time a little squarer with a top spread of 1.75m (5ft 9in). This is set under some trees and is looking for more effort than fence 1 but still giving the riders the chance to settle their nerves and affording them the opportunity to get their horses into a good rhythm.

3. *Bar* This is quite a big fence albeit straightforward. It is at maximum height and top spread coming around a left hand bend and one important point to bear in mind, particularly for the younger horses, is that it is where they will

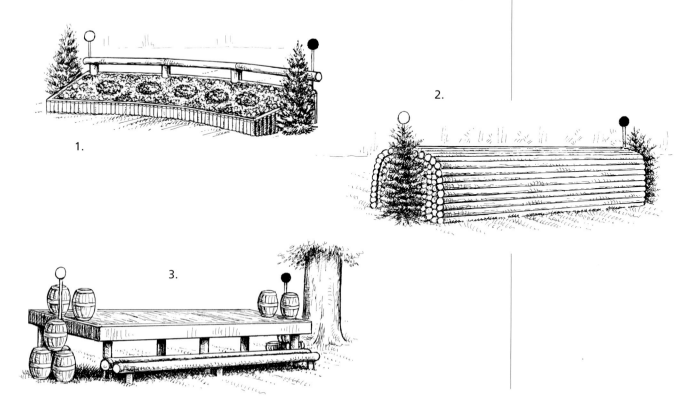

1.

2.

3.

see a lot of spectators and general activity for the first time since it is positioned close to the main public entrance.

4. abc ***Normandy Bank*** This is the first question on the course. Horses and riders will have used the first three fences to their advantage to get into the course and are now being asked to concentrate a little more. There is a step up of 1.13m (3ft 8in) and then a bounce over a log onto a down slope (kind to the horses without really changing the question) followed three strides later by a narrow triple brush. There is an alternative to 'B' and 'C', the alternative being a smaller log with a drop swinging around to another triple brush that is a little wider at 30cm (12in) than the one on the direct line. The three strides to the triple brush are set on a moving distance so that the riders have to trust their horses to be honest. We want horses to be thinking positively and confidently. The log itself is no more than 1.07m (3ft 6in) high but looks big from the bottom of the step and the question is one of power, athleticism, and confidence since the horses cannot see the landing after the log until late.

5. ***Oxer*** A maximum height and spread fence over a scoop that is 45cm (18in) deep; the base spread is 2.45m (8ft). A big fence incorporating the first 'ditch' on the course; the scoop has been created to serve as a warm up for the next fence.

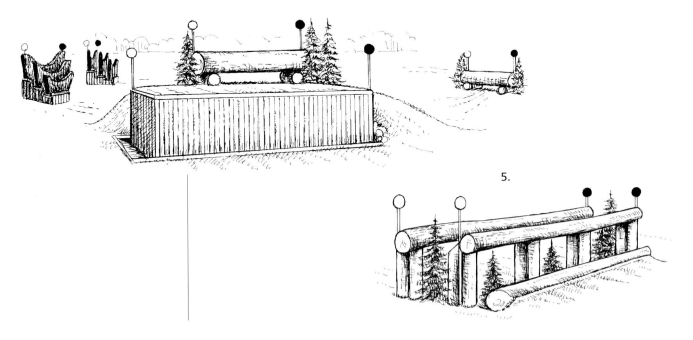

4. abc

5.

6. ***Ditch and Palisade*** Classic cross-country fence that looks imposing and rides really well provided it is ridden with commitment and in a positive frame of mind. The palisading is at 1.15m (3ft 9in), the brush is at 1.4m (4ft 7in), and the inside measurement of the ditch is 1.7m (5ft 7in). It is the first proper ditch on the course although the size of the take-off rail (20cm [8in] diameter) does take away some of the impact of the ditch as the horses approach. Both ends of the ditch are open in the unlikely event of a horse getting into it.

7/8. ab ***Dew Pond*** The first water fence on the course comprising a log with an alternative, bounce down into water off a step, followed by three strides up a slope to another log. 'A' is a shade less than 1.2m (3ft 11in) set so that the horses can see their landing area and the water over the top of it, the step down is 1.1m (3ft 7in), and the last element is at 1.2m (3ft 11in) on a good three strides that allows forward riding which is what is wanted coming up the slope. It is a fence that requires the rider to present his horse at the correct pace with plenty of impulsion; too slow will incur a refusal and if the horse is too

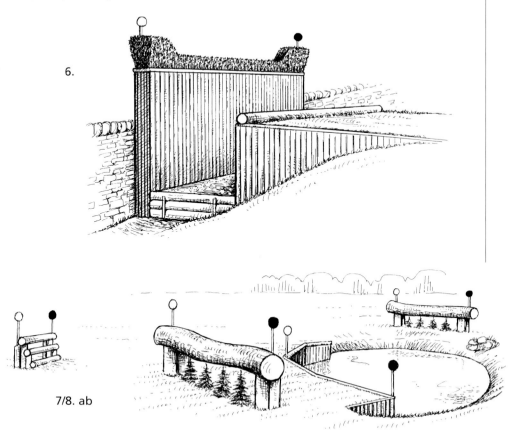

6.

7/8. ab

'long' it will not be in the right shape to jump the fence. The horse needs to be athletic and confident and must have trust in the rider. This fence is also in a busy public area and this too could impact on the horse's thinking. The long route is separately numbered so that it will flow and does not require the horses to be pulled around. It allows the riders to not have to jump the bounce element but it will take a lot longer.

9. ab **Offset Houses** Set on one stride, riders have to jump across these two elements at quite an angle left to right. There is deliberately no alternative because the feeling is that horses and riders at this level should be able to answer this question. It requires positive riding and honesty on the part of the horse not to run out. However, if the rider does not do a good job a run out is almost inevitable.

10. **Bullfinch** Very inviting 'let up' fence after the recent questions this also serves as the first 'stop' fence. There is a good apron on the front, the timber is at 1.15m (3ft 9in), the brush is at 1.4m (4ft 7in), and the bullfinch part is not

9. ab

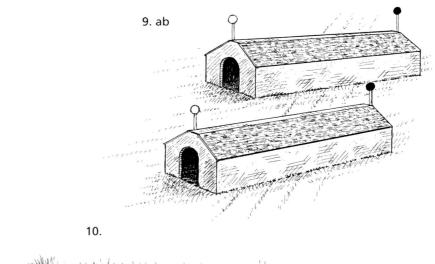

10.

very thick which is important because we are not looking for the horses to jump over it, we want them to brush through it.

11. **Sharks Teeth Oxer** This does not measure very big but it looks huge! It actually measures 1.13m (3ft 8in) at the back with a top spread of just 1.5m (4ft 11in) but what makes it look so big is that it is set on a slight down slope. The important point here is the front of the fence – the profile must be such that the horses are not sucked into the bottom of the fence and the top rail at the front must be high enough to get the horses in the air but not so high that it will flip horses over – the fence must be ascending and not have too much spread.

12. ab **Drop to House** A maximum drop off the platform onto sloping ground followed four strides later by a House at 1.15m (3ft 9in) set on top of a mound, also with a good drop onto ground that falls away. Riders need control off the drop and then to move to the second element which will jump big. From the top of the drop it looks a long way down and one or two horses may hesitate a

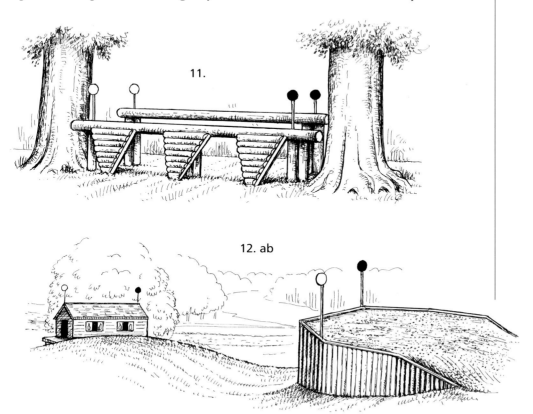

11.

12. ab

little. This therefore needs a fence judge who understands the rules clearly and also how horses organize themselves in order to negotiate this type of fence. It is essential that horses land on sloping ground at fences such as this so that their landing is comfortable, not jarring. If the ground is firm then some sand should be put down to act as a cushion.

13. *Table* Maximum height and top spread with a sloping front to make it inviting, this fence follows quickly on from the previous one. Anyone who found fence 12 a little intimidating should feel happy again after this.

14. ***Rails with Drop*** 1.1m (3ft 7in) post and rails off a right hand turn set at the top of a slope overlooking the second water fences. Even though this is some 70m (75yds) before the next fence how it is jumped will have a bearing on how the next fence jumps. There are three rails in the fence to help prevent horses being sucked into the bottom of it and therefore being too close and it requires control and neatness on the part of the horse. There is an alternative for those who may feel that their horses are a little casual in front.

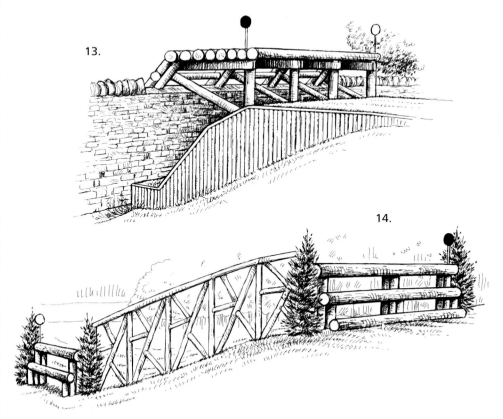

15/16. ***Bounce into Water*** Two sets of sloping rails bouncing into the water asking a different question to the first water fence. The first element is set at 1.13m (3ft 8in), the second element is at 1.02m (3ft 4in) and it also has a drop of 1.7m (5ft 7in). Apex to apex the distance is 4.11m (13ft 6in). The elements are numbered separately to allow for the long route to flow sympathetically for the horses. This is a difficult fence; horses have to be confident, bold, athletic, and in control, and the riders have to be committed and present their horses correctly. Anyone who was a little suspicious at the first water may well opt, very sensibly, for the long route here rather than risk a refusal. There is a long stretch of water facing the horses as they approach and so they have plenty of time to see it.

17. ab ***Step to Brush*** Coming out of the water after some 70m (75yds) the step is 90cm (3ft) high to a bounce over a brush fence made to look like a duck at 1.35m (4ft 5in). There is a long route for two reasons – firstly, in case of a refusal the riders do not have to go back into the water to try again, and secondly, for anyone not too confident about the question which does look

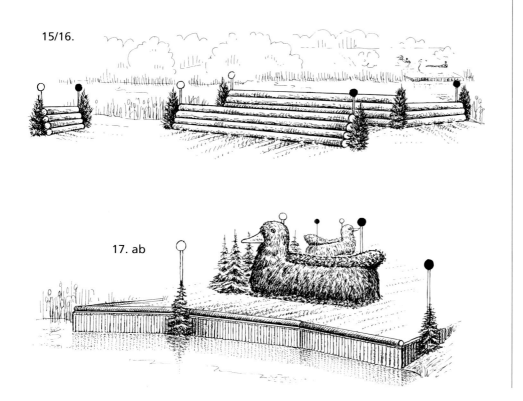

15/16.

17. ab

imposing. The top of the brush has 30cm (12in) through which the horses can brush and there is a good apron on the front. Again, different to the first water.

18. **Hammock** Maximum height and spread this is the second 'stop' fence on the course and it will also act as a confidence booster for anyone who was perhaps a little sticky at the water fences.

19. abc **Combination of Oxer to R/H Corner to R/H Corner** Back to work again with this sequence of fences which are all related to each other. The oxer is at 1.2m (3ft 11in) height with a top spread of 1.8m (5ft 11in), followed by a 75 degree corner at 1.2m (3ft 11in) after five or six strides on a gently curving left hand line, followed by another corner at the same height but only 70 degree after four strides on a gently curving left hand line. The long route is to jump the oxer and then take two smaller alternative corners to 'B' and 'C'. This is a

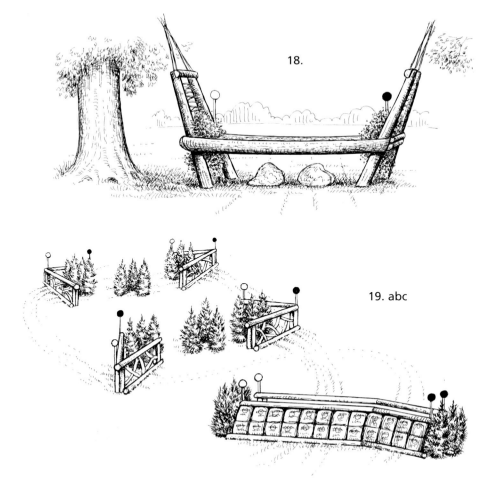

18.

19. abc

real rider fence and needs careful planning beforehand to ensure the correct line. The horses must be listening to the rider and be in control yet also be confident and bold. It would be very easy to run out at either of the corners, particularly the second one, if they are too bold and make up too much ground in the air. Neither of the corners is platformed but there are trees to prevent the horses jumping the corners where we do not want them to.

20. **Picnic Tables** A big fence at maximum height and spread which can act as a 'stop' fence and also as a confidence booster if anything went wrong at the previous combination.

21.abc **Rails to Bounce** Another combination with a set of rails at 1.1m (3ft 7in) off a slight downslope followed by five or six strides to a bounce of uprights with green aprons. The first element looks, and is, big enough given its positioning and then the riders have to curve left to the bounce. The bounce is set at 1.15m (3ft 9in) for the first part, and 1.2m (3ft 11in) for the second, with an apex to apex measurement of 4.6m (15ft 1in). Riders need to walk their line carefully in advance; they have the option of choosing the

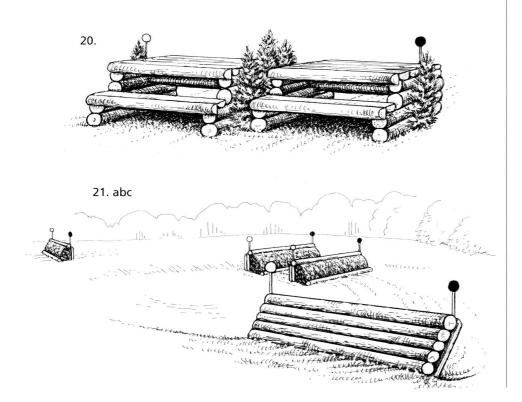

20.

21. abc

number of strides to suit their own horse because of the curving line. The horses need to be in control for the first element in order to do a good job at the bounce. There is only an alternative at the last element in the belief that horses at this level should be able to answer the question but should they have a stop at the second element of the bounce there is a means of jumping it without having to try both elements again.

22. ab/23. ***Coop to Pheasant Feeders*** Keeping the riders on their game and the horses paying attention, these two fences need thinking about but they are not too difficult although they do come only about 100m (110yds) after the previous fence. The coop is at 1.2m (3ft 11in) and has a top spread of 1.8m (5ft 11in) and then there is a turn to the two Pheasant Feeders on two strides, both going slightly downhill, angled right to left so that the line is jumping them both at an angle, and set at a height of 1.15m (3ft 9in). It is all too easy to run out at the second one. This is the opposite of Fence 9ab, the Offset Houses and the numbering is so that competitors can opt to jump the two feeders in an unrelated manner.

24. ***Brush Oxer*** Straightforward big fence with the brush at 1.4m (4ft 7in) high, the frame at 1.15m (3ft 9in), and a top spread of 1.75m (5ft 9in). There is

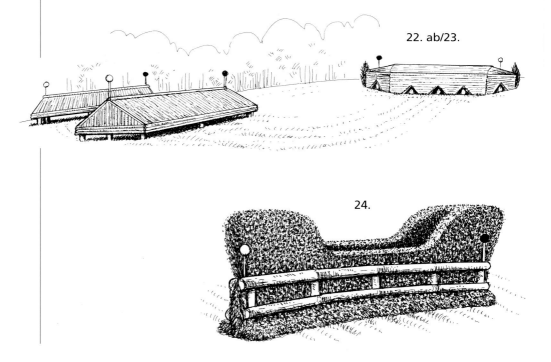

22. ab/23.

24.

a platform should anyone try to bank the fence and also a good groundline of brush. It comes after a long water crossing and is a bit of a let up fence that still needs jumping.

25. *Fort* At 1.2m (3ft 11in) high with a top spread of 1.65m (5ft 5in) it is set on a small flat area on a gentle, short uphill climb. It is important to not have much spread on this type of fence for fear of horses catching the back of it. This fence is here to break up the gallop to the next fence since we are getting towards the end of the course and also to keep the horse's mind on the job.

26. abcd *Double Corners* The last question where riders have the choice of two left hand corners or an upright to a right hand corner to another upright, i.e. a traditional 'W' shape fence. The fence is up to height, there are two strides between all elements, and the corners are platformed given that we are at the end of the course and horses may be tiring. It is good to keep horses and riders respecting the course and having to make decisions right up to the end. It would be very easy, and frustrating, to have a run out here.

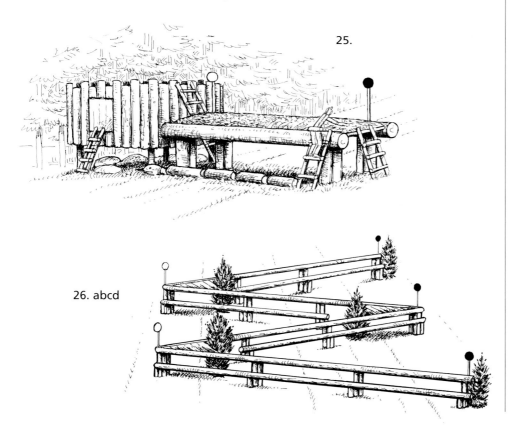

25.

26. abcd

27. *Feedrack* Two from home, maximum height and near maximum spread, straightforward fence with a solid top should any horse drop a leg or try to bank it for whatever reason.

28. *Arch* Off a right hand bend this fence is up to height but with only a small spread of 1.60m (5ft 3in) with an arch over the top. It has a good groundline with lots of 'bottom' so that horses will hopefully not get too deep and hit the fence.

No. of Jumping Efforts – 40

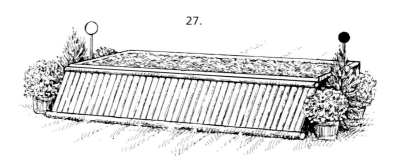

27.

28.

ASSESSMENT

Good variety of questions asked of horse and rider. The questions are more sophisticated than those for course B. There is a good opportunity for riders to get into the course before meeting the first question. The majority of questions asked in the middle of the course but riders still have to pay attention through to the end; decisions need to be made around the course based on how the horses are going. Eleven related fences/combinations. Horses and riders should benefit from the course and help to be prepared for the next level of competition.

Clearly the terrain impacts on the overall evaluation of a course. Courses A and C are gently undulating, course B has quite a steep a climb after fence 7 for about 100m (110yds). How to get horses up a slope is always a subject for debate. Obviously the preferred plan is to get a pull out of the way relatively early in a course (certainly not in the last third anyway) when the horses are still fresh and then give them a short breather at the top of the slope before asking them to jump anything. However, sometimes this is just not possible in which case it may be necessary to consider separating the start and finish to avoid an unnecessary climb.

If, having evaluated a course, there is concern that it is too strong or too easy, or the balance is not quite right, or there are too many similar questions, etc, then there is a need to discuss the concerns and try to resolve them before the competition takes place. The discussion needs to take place between the Course Designer, the Steward/Technical Delegate, and the Organizer in order to agree how to make any necessary corrections. All designers are guilty of over-doing some things, some times, and another pair of eyes is extremely valuable. The opinion of others should be welcomed, not resented.

Appendices

GUIDELINE DISTANCES

Some useful guidelines are set out below; they indicate inside measurements between elements and the slight variations depend on the level of competition, the type of fences, and the state of going. For competitions 1.05m to 1.15m in height the shorter distances are appropriate whereas for higher competitions the longer distances are more appropriate. What is important is that distances throughout all combinations must be consistent, e.g. do not mix long distances with short ones.

These will provide a guide and will jump well for most horses. Inevitably there are slight variations depending on the siting, the height and the type of fence and a designer needs to understand where and when to make any adjustments.

Cross-country

One stride	7.6m to 7.9m (25ft to 26ft)
Two strides	10.5m to 10.95m (34ft 6in to 36ft)
Three strides	14.6m to 15.8m (48ft to 52ft)
Steps up or down	2.78m (9ft)
Rail, bounce to step down	3.05m (10ft)
Step up, bounce to rail	2.78m (9ft)
Rail, one stride, step down	5.5m to 6m (18ft to 20ft)
Step down, one stride, rail	5.4m to 5.8m (18ft to 19ft)
Step up, one stride, rail	5.8m to 6m (19ft to 20ft)
Rail, one stride, step up	7.3m (24ft)
Sunken Road, one stride	6.55m to 6.86m (21ft 6in to 22ft 6in)
Sunken Road, two strides	9.15m to 9.45m (30ft to 31ft)
Coffin A to B	not less than 5.5m (18ft), (6.1m (20ft) on flat ground)
B to C	not less than 5.8m (19ft), (6.4m (21ft) on flat ground)
Bank – bounce	2.6m to 3.05m (8ft 6in to 10ft)

Show Jumping

Vertical to vertical

One stride	7.45m to 7.9m (24ft 6in to 26ft)
Two strides	10.5m to 10.95m (34ft 6in to 36ft)

Vertical to oxer

One stride	7.3m to 7.6m (24ft to 25ft)
Two strides	10.35m to 10.95m (34ft to 36ft)

Oxer to oxer (advanced level competitions)

One stride (rarely used)	7m to 7.3m (23ft to 24ft)
Two strides	10.35m to 10.65m (34ft to 35ft)

Ascending spread to ascending spread

One stride (rarely used)	6.85m to 7.45m (22ft 6in to 24ft 6in)
Two strides	10.2m to 10.8m (33ft 6in to 35ft 6in)

British Eventing Permitted Obstacles Dimensions 2003

CLASS	Intro	Pre Novice	Novice	Intermediate	Advanced
Max Height	90cm	1m	1.1m	1.15m	1.2m
With Height & Spread					
Max Top Spread	1m	1.1m	1.4m	1.6m	1.8m
Max Base Spread	1.5m	1.8m	2.1m	2.4m	2.7m
With Spread only					
Max Spread without height	1.2m	1.8m	2.8m	3.2m	3.6m
Drop Fences					
Max Drop	1.2m	1.4m	1.6m	1.8m	2m
Jumps into & out of water					
Max depth of water	20cm	20cm	30cm	30cm	35cm

FRANGIBLE FENCES

Over the years there has been much discussion about the actual fence construction techniques used in our sport and as part of the on-going look at safety and welfare British Eventing has introduced the use of Frangible Safety Pins into the UK.

Clearly in an ideal world there would not be horse falls and horses and riders would not be punished in this way for making mistakes. No-one wants to see horses falling and we all have a responsibility to work towards this goal.

It is considered that the high rotational falls are potentially the most dangerous to horses and riders and so any method of construction that helps to prevent this must be good. Many ideas have been put forward recently, all with a lot of good sense and much merit. Obviously we must be absolutely certain that any new construction technique is an improvement on what we already have and does not, in turn, create new problems. For this reason any new idea needs thorough research ahead of any introduction.

A brief explanation of the Frangible Pin system is as follows:

The pin is inserted to a specific depth into a metal sleeve (of a specific diameter and thickness) which has been inserted into the post to give the required height of the fence. The depth that the pin is inserted depends on the diameter of the rail to be supported and currently the permitted diameter of rails with this system is 16–25cm (6¼in–10in). The rail is roped such that the rail must be able to drop at least 40cm (15¾in) if and when the pin fractures and there must be no way that there is any chance of a horse having its leg trapped if and when the rail drops. There is a maximum weight of rail and any join between two rails has to be done in a specific, very simple, way.

This all sounds very complicated but in reality it is very simple. However, at the moment it is only suitable and appropriate for horses, not ponies, and should only be used by those familiar with the system to ensure correct application.

Imagine a horse is about to have a rotational fall. Extensive research has shown that when it gets to a certain point there is a significant downward force and if the rail can be got out of the way quickly (downwards by at least 40cm [15¾in]) the chances are greater that the rotational fall will not take place, thus reducing the potential crushing injuries. The pins are manufactured using special alloys and are designed to fracture when this designated downward force is applied – their characteristics need to be such that they are weather resistant and will not deteriorate if rubbed by a horse in the usual way of jumping.

Clearly they are only appropriate in certain types of fences at the moment but developmental work is going on to try to make them usable for bigger diameter rails.

It is important to state that this is just one of the ideas that has been put forward and that we all need to work continually to try to come up with ideas that improve safety. As stated earlier in the book, safety and welfare is, and must remain, the number one priority of all officials and we must appreciate the consequences of what we are designing and asking horses and riders to compete over.

Further details are available from British Eventing.

Badminton Horse Trials 2002 – 'Frangible Pin' installation.

CONCLUSION

In course designing and building there is always something new to learn and the sport moves quickly in many ways. Skill levels grow, building techniques develop, designers share experiences and hopefully improve, and whether our involvement is as a designer, builder, rider, or official, we all share a common goal and must work together for the benefit of the sport.

No doubt by the time this book is published there will have been some advances in improved course design and fence construction which we should all recognize as a positive development.